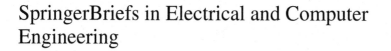

SpringerBriefs in Electrical and Computer Engineering

More information about this series at http://www.springer.com/series/10059

Bin Cao • Qinyu Zhang • Jon W. Mark

Cooperative Cognitive Radio Networking

System Model, Enabling Techniques, and Performance

 Springer

Bin Cao
School of Electronic and
    Information Engineering
Harbin Institute of Technology
    Shenzhen Graduate School
Shenzhen, Guangdong, China

Qinyu Zhang
School of Electronic and
    Information Engineering
Harbin Institute of Technology
    Shenzhen Graduate School
Shenzhen, Guangdong, China

Jon W. Mark
Department of Electrical
    and Computer Engineering
University of Waterloo
Waterloo, ON, Canada

ISSN 2191-8112   ISSN 2191-8120   (electronic)
SpringerBriefs in Electrical and Computer Engineering
ISBN 978-3-319-32879-9   ISBN 978-3-319-32881-2   (cBook)
DOI 10.1007/978-3-319-32881-2

Library of Congress Control Number: 2016937362

Printed on acid-free paper

This Springer imprint is published by Springer Nature
The registered company is Springer International Publishing AG Switzerland

Preface

With the explosive proliferation of wireless services and applications, such as vehicular ad hoc networks, smart grid, and Internet of Things (IoT), the demand for radio spectrum has been skyrocketing. Since the amount of usable radio spectrum is finite, frequency bands and their usage are strictly managed and enforced by governmental organizations. Under this regulatory enforcement, spectrum is statically and exclusively allocated to dedicated networks on a license basis, i.e., only primary users (PUs) can access the assigned spectrum.

Although interferences among different networks and devices can be efficiently coordinated by using fixed spectrum allocation, this policy causes significant spectral underutilization as measured and found in practical application environments. Besides the physical scarcity of radio spectrum, spectral underutilization also results in artificial scarcity. Moreover, the considerable new wireless applications and services along with emerging diverse wireless networking architectures, e.g., heterogeneous networks, have been hampered by the inefficient allocation of radio spectrum. This leads to the contradiction that proliferation of wireless applications and services starves for spectrum, while large portions of spectrum are unused most of the time by PUs.

As mentioned above, legacy command-and-control spectrum allocation leads to significant spectral underutilization and inefficiency owing to the sporadic use of spectrum. We are facing a contradictory situation that the practically explorable radio spectrum is physically scarce, while the assigned spectrum is significantly underutilized. The inefficient use of spectrum has hindered the sustainable revolution of wireless technologies and has been an urgent bottleneck to further proliferation of wireless services and applications.

In addition, electric power consumed in wireless communications and networks is exponentially increasing. The associated energy consumption is emitting non-negligible amount of greenhouse gas, resulting in environmental impacts, e.g., global warming. On the other hand, the implementation to fulfill advanced signal and information processing functionalities, such as multiple-input and multiple-output (MIMO) technologies, consumes considerable energy and requires a large physical size, and this problem is particularly critical to the wireless devices and networks with strict size/cost and energy/power limits, such as handheld terminals and wireless sensor networks.

Radio spectrum underutilization and energy inefficiency become urgent bottleneck problems to the sustainable development of wireless technologies. The research philosophies of wireless communications have been shifted from balancing reliability-efficiency tradeoff in the link level to seeking spectrum-energy efficiency in the network level. Global spectrum-energy efficient designs attract significant attention to improving utilization and efficiency, wherein cognitive radio networks and energy-efficient resource allocation are of particular interests.

To achieve sustainable development of wireless industry and to support diverse wireless applications and services, new system design methodologies and solutions with high spectral efficiency are imperative. Dynamic spectrum access (DSA) networking is an attempt to improve spectral efficiency by dynamical and opportunistic making use of licensed spectrum by SUs, and the network consisting of SUs is a secondary network. This novel and promising networking architecture has a lot of technical and business opportunities to exploit, such as the ongoing TV white spaces and Wi-Fi 2.0.

Since PUs are authorized to use licensed spectrum, the most important issue in DSA design is to avoid interferences from SUs to PUs if SUs temporarily access the dedicated licensed spectrum, i.e., SUs transmission must be transparent to PUs. One feasible methodology for SUs avoiding interferences to PUs is SUs transmit their signals opportunistically by sensing the availability of unused spectrum bands, i.e., spectrum holes. These spectrum holes indicate the absence of PUs usage either in frequency, in time, or in space. Besides using spectrum when PUs are absent or inactive, SUs can also share licensed spectrum with active PUs as long as PUs' transmission is not interfered by SUs. An SU with such capabilities is also named as a *cognitive radio* (CR), and the network consisting of CRs is called *cognitive radio networking* (CRN).

This monograph aims to address the energy-efficient design and resource management in cognitive radio networks, with the emphasis on system modeling, key enabling technologies in physical layer and medium access control (MAC) layer. To this end, we firstly provide a systematic study on active cooperation between primary users and secondary users, i.e., cooperative cognitive radio networking (CCRN), followed by the discussions on research issues and challenges in designing spectrum-energy efficient CCRN in Chap. 1. As an effort to shed light on the design of spectrum-energy efficient CCRN, we model the CCRN based on orthogonal modulation in Chap. 2, wherein system model, two-phase and three-phase cooperation frameworks, and performance analysis will be covered. In Chap. 3, by exploiting orthogonally dual-polarized antennas (ODPAs) in CCRN, we detail ODPA-based CCRN paradigm, key enabling techniques, and resource management issues. To further enhance the system performance of CCRN, we introduce the optimal communication strategies for both primary and secondary users in Chap. 4. Conclusion remarks and future potential research issues are listed in Chap. 5.

We hope this monograph can provide the readers an entrée into this very active field, aiming at covering the state-of-the-art aspects of analysis and design of energy-efficient CCRN. We would like to thank the supports from National Natural Sciences Foundation of China (NSFC) under Grant No. 61032003 and 61501211, Postdoctoral Science Foundation of China under Grant No. 2014M560263, Natural Science Foundation of Jiangxi under Grant No. 20151BAB217001 and 20151BAB217018, and S&T Foundation of Jingdezhen.

Shenzhen, Guangdong, China   Bin Cao
Shenzhen, Guangdong, China   Qinyu Zhang
Waterloo, ON, Canada   Jon W. Mark
March 2016

Contents

Chapter 1
Introduction

Abstract This chapter presents fundamental issues of cognitive radio networks (CRNs). In the first section, background of why using CRN and some conceptual descriptions about CRNs will be given, wherein interleave-type CRN, underlay-type CRN, and overlay-type CRN are detailed. In addition, current research topics, study interests, as well as challenging issues of each CRN type will be discussed. In the second section, user cooperation in CRNs will be illustrated, wherein cooperation between primary users (PUs) and secondary users (SUs), and cooperation among SUs, are covered, respectively. As the mainline of this monograph, we study active cooperation between PUs and SUs, which is referred to as cooperative cognitive radio networks (CCRN) in the third section. In this part, cooperation frameworks, transmission technologies, and protocol designs will be reviewed. To shed light on this very topic, the associated research directions and challenges will be discussed. Finally, state-of-the-art of CCRN is listed for the readers to have better understanding about CCRN.

1.1 Cognitive Radio Networks

We have witnessed that wireless communications enjoys a big success both in engineering and academia aspects over the last 30 years, and becomes the fastest developing information technologies (IT). Until late 1990s, wireless communications has been associated with cellular telephony, such as GSM and TDMA, as this is the biggest market segment of voice and instant message services, and has played an important role in people's daily lives. From 2000s, with the extensive use of wireless computer networks, wireless communications has changed not only lives but also working habits and mobility of people. Wireless sensor networks (WSNs) monitor factories and environments, wireless links replace the cables between computers and keyboards, and wireless positioning systems, such as Global Positioning System (GPS) and BeiDou Navigation Satellite System (BDS), help with navigating drives to their intended destinations, radio frequency identification (RFID) technologies monitor the location of trucks that have goods identified by wireless radio frequency tags.

© The Author(s) 2016
B. Cao et al., *Cooperative Cognitive Radio Networking*, SpringerBriefs in Electrical and Computer Engineering, DOI 10.1007/978-3-319-32881-2_1

In recent decade, we again are shocked by the explosive proliferation of applications in wireless communications and networking, e.g., vehicular ad hoc networks (VANETs), internet of things (IoTs), social networks, location-based services (LBSs), and smart grid. State-of-the-art technologies associated with products and services are changing our life styles from various aspects. While we are now enjoying the remarkable improvement in quality of service (QoS) provided by the fourth generation (4G) mobile communications networks, i.e., LTE-Advanced and WiMAX-Advanced, people from both academia and industry are seeking for the design of the next generation (xG) wireless networks, among them Wireless Vision 2020 plan towards 5G standard is notable .

To support sustainable development of wireless communications and networks, the demand for radio spectrum has been skyrocketing. Since the amount of usable spectrum is finite, frequency bands and their usage are strictly managed and enforced by government regulators. In the past decade, information and communication technologies (ICT) industry has turned into a highly competitive industry where companies are competing to buy valuable spectrum from governmental bodies, i.e., spectrum auction. In such a way, the winning telecommunications operators in spectrum bidding obtain the rights, also known as licenses, to transmit signals over specific bands of the electromagnetic spectrum and to conduct their communications services. With more services providers in the mobile industry, the competition during spectrum auctions has increased due to more demand from consumers and physical scarcity in available radio spectrum. In early 2013, for commercializing 4G networks, the United Kingdom government performed a spectrum auction for 250 MHz of spectrum (equivalent to two-thirds of the entire 3G spectrum already in use) in the 800 MHz and 2.6 GHz bands. In total, five bidders have committed to pay 2.34 billion UK pounds for the right to use the frequencies for 4G services. According to a report from the Ministry of Industry and Informationization of China, China's total spectrum demand in wireless communications is about 1 GHz by 2015. However, 547 MHz of spectrum has already been assigned for IMT systems, leading to 420 MHz of spectrum scarcity, if no new spectrum is exploited. Figure 1.1 is the radio frequency allocation chart of Hongkong Special Administrative Region in China, showing that most of the easily usable spectrum nearly runs out.

On the one hand, we are now facing a challenging issue that physical spectrum scarcity hinders the further evolution of wireless communications. On the other hand, investigations and experiments on spectrum utilization have clearly shown that most of the allocated spectrum is considerably underutilized, as shown in Fig. 1.2. Since the spectrum is exclusively assigned to dedicated users, termed as licensed or primary users (PUs), other users cannot access to the spectrum even if it is unused. This also imposes us a conflict between spectrum scarcity and spectrum underutilization, which further exacerbates the situation of spectrum shortage. New wireless applications and services along with emerging diverse wireless networking architectures, e.g., heterogeneous networks, have been severely hampered by this inefficient fixed spectrum allocation.

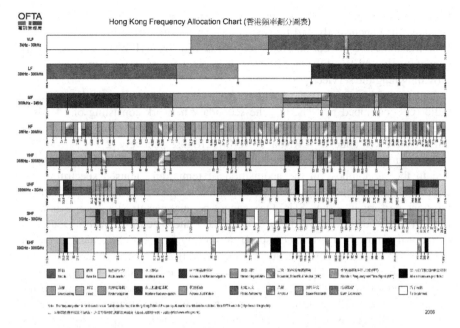

Fig. 1.1 Radio frequency allocation chart of Hongkong Special Administrative Region

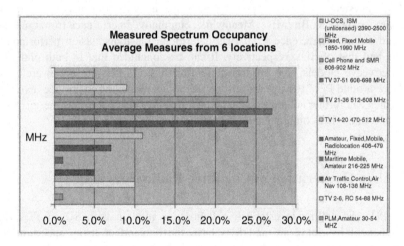

Fig. 1.2 Measured spectrum utilization of some typical bands (average measures from six locations)

In order to address spectrum underutilization, dynamic spectrum access based schemes, which allow unlicensed users (or referred as secondary users (SUs)) to share or reuse licensed bands without interfering with PUs, have attracted significant attention [1–7]. By enabling dynamic spectrum access, SUs are allowed to (i) dynamically sense the surrounding electromagnetic environments and opportunistically access to temporally unused spectral bands, e.g., spectrum sensing based systems, (ii) concurrently and transparently transmit in licensed bands as long as PUs' transmissions are not interfered, e.g., ultra-wide band (UWB) systems, or (iii) cooperatively and trustfully negotiate with PUs for transmission opportunities by providing tangible services [2].

One promising technology to achieve dynamic spectrum access is the cognitive radio (CR) equipped by SUs, i.e., the secondary network consisting of CRs is a cognitive radio network (CRN). In this context, these three paradigms are known as interweave CRN, underlay CRN, and overlay CRN, respectively [2].

In an interweave CRN, communication durations of SUs are highly unstable due to the randomness in the acquisition of temporally unused spectrum, and transmission terminations when PUs reclaim the spectrum degrade SUs' transmissions performance, i.e., SUs' QoS cannot be guaranteed. Moreover, SUs need to exert more power and overhead to attain accurate spectrum sensing. Detecting temporarily unused spectrum, i.e., spectrum sensing, is one of the critical elements in CRN design, and performing an accurate spectrum sensing is a challenging task.

In an underlay CRN, like the impulse radio ultra-wide band (IR-UWB) wireless communication systems, since SUs' communications cannot produce any harmful interference to PUs' communications, the interference temperature model is imposed on SUs' transmit power to avoid interference to PUs' transmissions [2]; this limits SUs to short-range communications. Such as in impulse-radio ultra-wideband (IR-UWB) systems, the allowed power spectrum density for transmission should be strictly below -41.3 dBm/MHz. Meanwhile, cumulative interference from multiple SUs and strong interference from PUs deteriorate the transmission performance of PUs and that of SUs, respectively. It can be concluded that in both underlay and interweave models, SUs are transparent to PUs, i.e, PUs are not aware SUs' existence around PUs. However, user cooperation is promising to provide diversity gain which can further enhance the resource utilization, like spectrum, energy, time, and space, and other communication resources.

1.2   User Cooperation in Cognitive Radio Networks

Smart PUs are willing to establish collaborative relation with SUs if PUs can obtain benefit by doing so, and SUs have the demand of using spectrum for transmission, which forms the overlay-type CRN. In response to the challenging issues in spectrum sensing based CRN, an alternative promising approach for SUs to gain transmission opportunities using licensed spectrum is to provide tangible services to PUs such that the PU can transmit its traffic to its intended destination with greater reliability, e.g., faster and with satisfactory quality. If the SU can provide such service, the PU would be happy to yield a fraction of its licensed spectrum for the SU to use, so long the SU's transmission does not interfere with the PU's transmission. In this way, the PU and the SU mutually benefit from user cooperation, i.e., the PU and the SU cooperate to achieve mutual benefit.

In this section, we talk about two types of user cooperation fashions in a cognitive radio network, namely, cooperation among SUs, and cooperation between SUs and PUs. In the first terminology, SUs cooperate to have more accurate spectrum sensing performance while with less computation complexity, while SUs cooperate with PUs to have more smooth transmission status with decreasing signaling overhead.

1.2.1 Cooperation Among Secondary Users

In interleave based CRN, since SUs need to occupy the licensed bands of PUs, SUs must detect the presence of PUs in a very quick fashion and must release the occupied bands as long as PUs are active. In this regard, one of the most challenging issues that confronts this concept is how SUs detect the presence of PUs. Detecting temporarily unused spectrum, i.e., spectrum sensing, is one of the critical elements in CRN design, and performing an accurate spectrum sensing is a challenging task. The sensing performance highly depends on channel quality and sensing time. For example, in low signal-to-noise ratio (SNR) environments, the effect of SNR walls deteriorate and limit sensing ability, which leads to interferences to PUs from SUs because of missing detection of spectrum holes.

Another notable challenging task of implementing spectrum sensing is the hidden terminal problem, which frequently occurs when SUs are shadowed in severe multipath fading environments or inside buildings with high penetration loss, while PUs are operating in the vicinity. Due to this terrible hidden terminal phenomenon, an SU may fail to detect the presence of PUs and then will access the licensed spectrum and cause severe co-channel interference to PUs. In order to address this issue in CRNs, multiple SUs can cooperate to perform spectrum sensing.

Cooperative communications based spectrum sensing in CRN, i.e., cooperation among SUs, has an analogy to a distributed decision in WSNs, where every single node makes a local decision and those decision results are sent to a fusion center so as to achieve a final decision according to associated fusion principles.

Basically, cooperation among SUs, i.e., cooperative spectrum sensing can be performed as following steps. In the first step, each individual SU independently detects the spectrum usage in its surrounding area by using some spectrum sensing algorithms (such as energy detection, matched filter, cyclostationary detection, and wavelet detection, etc), and makes its own decision whether PU is on or off. In the second step, all involved SUs send their binary decisions to a common fusion center. In the final step, this fusion center makes a final decision based on SUs' results according to some rules, such as hard decision fusion, soft decision fusion, and quantized decision fusion, etc. A classical block diagram of cooperative spectrum sensing architecture is shown in Fig. 1.3.

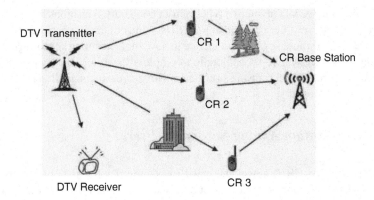

Fig. 1.3 Cooperative spectrum sensing in secondary networks. CR 1 is shadowed over the reporting channel and CR 3 is shadowed over the sensing channel [14]

1.2.2   Cooperation Between Secondary Users and Primary Users

Detecting temporarily unused spectrum, i.e., spectrum sensing, is one of the critical elements in CRN design, and performing an accurate spectrum sensing is a challenging task. The sensing performance highly depends on channel quality and sensing time. For example, in low signal-to-noise ratio (SNR) environments, the effect of SNR walls deteriorate and limit sensing ability, which leads to interferences to PUs from SUs because of missing detection of spectrum holes. Although some sensing schemes are with satisfied performance [14], e.g., cooperative sensing among SUs, they are with high complexity for implementation. Furthermore, spectrum sensing based methodology cannot fully guarantee SUs transmission performance even though SUs are with accurate sensing ability, since an SU must abandon its transmission as soon as it detects the presence of the PU, and sense another spectrum hole for transmission. If none is available, then the SU has to discontinue the transmission. The main reason leads to this issue is due to PUs are unaware of SUs demand and existence, which indicates that there is no collaborative relation between PUs and SUs.

Although some sensing schemes are with satisfied performance [14], e.g., cooperative sensing among SUs, they are with high complexity for implementation. Furthermore, spectrum sensing based methodology cannot fully guarantee SUs transmission performance even though SUs are with accurate sensing ability, since an SU must abandon its transmission as soon as it detects the presence of the PU, and sense another spectrum hole for transmission. If none is available, then the SU has to discontinue the transmission. The main reason leads to this issue is due to PUs are unaware of SUs demand and existence, which indicates that there is no collaborative relation between PUs and SUs.

Smart PUs are willing to establish collaborative relation with SUs if PUs can obtain benefit by doing so, and SUs have the demand of using spectrum for transmission. In response to the challenging issues in spectrum sensing based CRN, an alternative promising approach for SUs to gain transmission opportunities using licensed spectrum is to provide tangible services to PUs such that the PU can transmit its traffic to its intended destination with greater reliability, e.g., faster and with satisfactory quality. If the SU can provide such service, the PU would be happy to yield a fraction of its licensed spectrum for the SU to use, so long the SU's transmission does not interfere with the PU's transmission. In this way, the PU and the SU mutually benefit from user cooperation, i.e., the PU and the SU cooperate to achieve mutual benefit. This is called *cooperative communications* between PUs and SUs in cooperative CRN (CCRN).

Another promising approach for SUs to gain transmission opportunity using licensed spectrum is to lease PUs' temporarily unused spectrum, e.g., SUs lease spectrum from inactive PUs by paying leasing cost. For an inactive PU, it is willing to lease unused spectrum to SUs for monetary reward, since the PU pays financial cost to gain the licensed spectrum. In this way, the PU and the SU are both motivated to participate into *spectrum leasing* in CCRN. Due to signal attenuation, one spectrum band can be reused by different SUs if they are far away from each other, i.e., the same with spectrum allocation in cellular networks. As an economy model, one significant issue in spectrum leasing is pricing schemes, wherein auctions are extensively investigated [12].

1.3 Active Cooperation Between Primary and Secondary Users

Although the interweave CRN appears to have no obligation on the part of SUs to PUs, durations of secondary transmission opportunities are highly random due to the randomness in the acquisition of idle spectral bands via spectrum sensing. Meanwhile, the termination of transmissions when the associated primary transmission is sensed degrades SUs' transmission performance. On the one hand, since accurate spectrum sensing is challenging, an SU may need to exert high energy levels, time and communications overhead to achieve the interweave CRN. On the other hand, through negotiation, an SU is fully aware of its obligation when it enters the cooperation agreement with a PU in an overlay CRN. Therefore, secondary transmission opportunities are dedicated rather than random in the context of cooperation between PUs and SUs. In an overlay CRN, an SU can obtain transmission opportunities from the PU in exchange for services provided to the PU, e.g., by relaying the PU's traffic [8–10], or by leasing a spectral band from an inactive PU [11].

The focus of this section is on the gain in transmission opportunities for SUs via overlay CRNs. In this scenario, the PU is the leader in the partnership, while the SU is the follower. Hence the PU can select the most appropriate SU for cooperative communications. In the case of spectrum leasing by an SU or a group of SUs, there can be cooperative communications among SUs. Traditionally, spectrum leasing involves pricing; some leasing strategies are discussed in [11–13]. This section assumes that an acceptable leasing price has been set, and focuses on structuring and analyzing cooperation among SUs.

User cooperation to improve spectrum efficiency and utilization in an overlay CRN is also termed cooperative cognitive radio networking (CCRN) [8–10, 14]. This article introduces two cooperation frameworks in CCRN to improve spectrum efficiency and utilization. Firstly, cooperation between active PUs and SUs [10], and the traditional spectrum leasing between SUs and inactive PUs are discussed. Cooperative communications between SUs and active PUs using a quadrature signalling scheme by leveraging two orthogonal channels in the two-dimensional modulation is then described, and cooperation among SUs in the mode of cooperative spectrum leasing and transmission deployment follows. Cooperation performance of each framework is evaluated and discussed by maximizing a weighted sum throughput problem under certain power and throughout constraints, and simulations show the feasibility of the proposed frameworks. Finally, concluding remarks and thoughts on future possible research issues end this article.

1.3.1 Cooperative Communications and Spectrum Leasing

In CCRN, cooperative communications and spectrum leasing between PUs and SUs have attracted considerable attention and have been extensively investigated separately in recent years [8–10, 13, 14]. In this section, some state-of-the-art work associated with challenging issues arising in these two areas are reviewed.

1.3.1.1 Cooperative Communications Between PUs and SUs

To gain transmission opportunities in CCRN, one or more SUs can act as relaying nodes for a PU. In its role as a relaying node, an SU can serve to provide a multi-hop relay service, or to provide an additional transmission path to the destination of the PU, and the PU yields the licensed spectrum to its relaying SU for a fraction of the time in return. In this case, the primary link can be established with the help of multi-hop transmissions, or the receiving PU can obtain diversity gain and enhance reception by appropriately combining the signals from the direct and relayed paths. Therefore, spectrum efficiency and utilization are significantly improved.

Relaying is a signal forwarding service, which can be implemented, for example, by using an amplify-and-forward (AF) or a decode-and-forward (DF) mode. Whether an SU is able to offer an effective relaying service to a PU depends on the quality of the propagation channels between the PU and SU, and the PU's destination and SU. Hence the PU would select the SU that can serve as the most effective relay amongst a set of candidate SUs. Since the PU is entitled to the right of controlling the assigned spectrum, it has the freedom to select the most appropriate SU as its cooperator. To gain transmission opportunities via relaying the primary traffic, the SU uses additional transmit power to forward the PU's information, so that the SU needs to evaluate its gain and cost in the cooperation process, indicating that the SU would also select the most appropriate PU as its cooperator.

There are different ways in which PUs and SUs can perform cooperative communications [8, 9], and three models are listed here as examples:

- Three-phase time division multiple access (TDMA) based cooperation: The PU transmits the primary traffic to its intended destination and the selected relaying SU or SUs in the first phase; the SU or SUs relay the PU's data in the second phase; and the SU or SUs transmit their own signals in the third phase. The most critical parameters of this scheme are the optimal time duration in each phase for both the PU and SUs, and the optimal allocation of transmit power levels of the PU and SUs for energy-efficient transmissions. Furthermore, the multi-user cooperation in the time domain, i.e., phases 2 and 3, may result in high communication overhead and collisions which degrade the cooperation performance of CCRN.
- Two-phase frequency division multiple access (FDMA) based cooperation: The PU uses a fraction of its licensed bandwidth for relay transmissions with an SU in a two-phase manner, and exclusively allocates the remaining bandwidth for the SU to address secondary transmissions. Since the SU can continuously transmit its own signal on a dedicated licensed band, the achievable throughput of the SU can be guaranteed. However, the role of CCRN is to guarantee the PU's transmission performance, while achieving an elastic throughput for the SU according to the varying wireless environment. Therefore, given the bounded system capacity, the PU may not willing to yield dedicated spectral bands to ensure the SU's throughput requirements.
- Two-phase space division multiple access (SDMA) based cooperation: The SU exploits multiple antennas to enable multiple-input-multiple-output (MIMO) capabilities, such as spatial beamforming, to avoid interference to the PU and interference to other SUs in CCRN. The MIMO-CCRN framework is proposed to allow the SU to use the degrees of freedom provided by the MIMO system to concurrently relay the primary traffic and transmit its own data at the cost of complicated antenna operation and hardware requirements. Since MIMO technology is not yet widely and readily deployed due to hardware and cost constraints of user devices or SU-type devices, this scheme is currently of less interest.

1.3.1.2   Spectrum Leasing Between PUs and SUs

PUs typically obtain licenses to operate wireless services, such as cellular networks, by paying spectrum regulators. In this context, one approach to attain CCRN is spectrum leasing that adopts pricing based incentives to stimulate PUs to lease their temporarily unused spectrum to SUs in return for financial reward. In a spectrum leasing (or spectrum trading) model, one challenging issue is the pricing problem, e.g., since spectrum providers or PUs compete with each other to lease their licensed spectrum to SUs, and SUs compete with each other to lease spectrum from PUs, pricing is complicated.

To achieve an efficient dynamic spectrum leasing protocol between PUs and SUs, some models based on economics have been introduced so as to maximize PUs' revenue and SUs' satisfaction. However, there is a trade-off between these two goals. One particular form of spectrum leasing is via auctioning which is widely used in providing efficient distribution and allocation of scarce resources [11]. By introducing an auction mechanism into CCRN, PUs can maximize their revenue through dynamical and competitive pricing based on SUs spectrum usage demand. Due to the unique characteristics of the radio spectrum, one spectral band can be reused in different areas because of signal attenuation during propagation, which means there may be multiple winners after an auction.

Another requirement in the spectrum auction model is that the auction mechanism should be quickly conducted to enable on-demand and instantaneous services of SUs, which means SUs' bidding should be processed immediately by PUs (or special brokers). Due to the complicated relationship between bidders and auctioneers, and the unique interference-limited characteristics, the overhead of auctions should be fully considered in the auction mechanism design. Otherwise, SUs are not willing to participate in auctions due to high overhead.

1.3.1.3 Challenging Issues

As aforementioned, the spectrum efficiency/utilization and transmission performance can be significantly improved by introducing user cooperation into CCRN. However, several challenging issues should be tackled.

- *Motivation for cooperation*: To establish a collaborative relationship between different users, there must be motivation for partners to cooperate. Therefore, stimulating motivation to cooperate is a prerequisite to establishing cooperation. Meanwhile, performance metrics to evaluate cooperation performance should be addressed. In this chapter, two simple cooperation frameworks which are capable of stimulating both active and inactive PUs, and SUs to participate in cooperation and achieve mutual benefit in CCRN are structured. To generalize the evaluation metrics in terms of throughput maximization, a weighted sum throughput optimization problem is formulated and solved for each framework.
- *Physical-layer issues*: To fulfill ideal cooperative communications, advanced physical-layer signal processing capabilities are indispensable. Synchronization between cooperators, orthogonal transmissions and relaying for non-interfering communications must be carefully designed in a complicated time-varying transmission environment. To this end, we present a simple and efficient orthogonal signaling scheme based on quadrature amplitude modulation (QAM) to attain orthogonal transmissions between cooperators.

- *MAC-layer issues*: To achieve satisfactory cooperation performance, finding the most efficient and effective cooperator is a challenging issue. The medium access control (MAC) layer protocol must coordinate the cooperator selection and management simply and efficiently. High communication overhead and energy consumption during multi-user coordination degrade the cooperation performance of CCRN. In addition, any potential interference resulting from the multi-user coordination should be avoided. To tackle this issue, a novel FDMA based multi-user coordination scheme is proposed in this article.

- *Co-channel interference issues*: To avoid co-channel interference when spectrum leasing is exploited, a given spectral band is reused in different areas similarly to the spectrum reuse in a cellular network. In this vein, spectrum reuse in CCRN is limited spatially. A promising method to further efficiently use temporarily unused spectrum is to cooperatively reuse the same spectrum when multiple SUs are within interfering range. We propose a cooperative spectrum leasing scheme for neighboring SUs to lease one spectral band together from an inactive PU, and to perform cooperative communications with each other. In this context, co-channel interference among SUs is avoided, and the transmission performance of secondary links and spectrum efficiency are further improved.

System Architecture and Description

In this section, an overview of the system architecture for CCRN is presented. The primary and secondary networks in infrastructure mode are designed as shown in Fig. 1.4: In CCRN, we consider the scenario consisting of one PN with a primary base station (PBS) and multiple PUs, and one SN with a secondary base station (SBS) and multiple SUs. In the PN, the PBS allocates network resources to PUs, e.g., spectrum, time slots, etc., so that PUs can access the spectral bands without interfering with each other. In addition, we consider that there is a dedicated control channel between the PBS and SBS. We take into account that both active and inactive PUs coexist in the PN. Each SU equipped with a single CR has the knowledge of channel state information (CSI) in terms of receiving signal-to-noise ratios (SNRs) of their interesting users, i.e., PUs and SUs can estimate and share CSI of the primary and secondary links; the details of this issue will be discussed in this section. Moreover, SUs are assumed to have advanced signal processing functions, such as adaptive modulation, coding and frequency agility.

To deploy user cooperation with high spectrum efficiency in CCRN, the foremost issue is how to exploit the available degrees of freedom in the wireless network, e.g., time, space, coding, modulation, etc, and how to efficiently manage and use these degrees of freedom is critical to the CCRN design and implementation. Another issue that should be taken into consideration is how to stimulate motivation for PUs and SUs to cooperate in CCRN.

As shown in Fig. 1.4, we discuss two types of user cooperation models in CCRN. The first one is the cooperative communications between SUs and active PUs, in which an SU relays a PU's packets and obtains a transmission opportunity as a

Fig. 1.4 Networking
architectures in CCRN

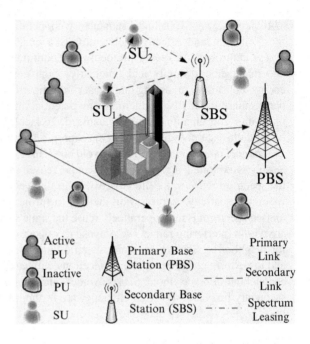

reward. The other model is the cooperative spectrum leasing among SUs in which
SUs together lease an unoccupied licensed band from an inactive PU and establish
cooperative communications with each other.

In the SU-PU cooperative communications model, an SU competes with other
SUs for cooperation with an active PU. To improve the opportunity of being selected
by the PU, an SU must optimize some performance metrics to enhance the PU's
transmission as much as possible because the PU would like to select the SU that
can offer the highest gain by cooperation. In this context, the SU uses additional
transmit power to forward the PU's traffic if it is selected as the relaying node by the
PU. In the SU-SU cooperative spectrum leasing model, an SU would like to work
with other SUs cooperatively to improve transmission performance and reduce the
cost for spectrum leasing as much as possible. To fulfill this objective, an SU and
its partners jointly optimize some metrics to attain transmissions in an economical
manner and with satisfactory performance.

Cooperative Communications Between SUs and Active PUs

In this section, a novel cross-layer cooperative communication framework based
on quadrature signaling for an SU cooperating with one active PU in CCRN is
discussed [10].

Cooperation Motivation

Consider an uplink scenario, when a PU cannot communicate with the PBS efficiently, e.g., if the PU is located far away from the PBS or the link between the PU and PBS is blocked by buildings as shown in Fig. 1.4, the PU may wish to select an appropriate SU to help relay the signal by yielding a fraction of its licensed spectrum for the SU to use. By selecting the SU as a relaying node, the PU can communicate with the PBS reliably and/or efficiently, e.g., the SU can provide multi-hop service to the PU as illustrated in Fig. 1.4. By relaying the PU's traffic, the SU can gain the opportunity of using the PU's licensed spectrum as a reward. Thus, the PU and the SU cooperate to achieve mutual benefit.

MAC-Layer Multi-User Coordination

To find an appropriate SU as the relaying node of the PU, a simple and interference-free MAC-layer multi-user coordination scheme can be used as shown in Fig. 1.5a. Firstly, the PBS informs the SBS of the cooperation information, including the operating spectral band of the PU and the SNR of the primary link. Secondly, the SBS broadcasts the received cooperation information to SUs via an unlicensed spectral band, e.g., an ISM band. Meanwhile, the SBS allocates a sub-band to each SU for replying to the PU. For example, the SBS divides the PU's bandwidth into N non-overlapping segments if there are N SUs managed by the SBS, i.e., each SU can obtain an orthogonal sub-band for transmitting its response to the PU; therefore, there is no interference among SUs when responding to the PU. After that, SUs switch to the current operating band of the PU. To allow SUs to measure the CSI of the link between the PU and SUs, the PU broadcasts a cooperation request in its operating channel. Thirdly, after getting the PU's information and the associated SNRs, each SU designs an optimum transmission and relaying strategy. If the cooperation can achieve mutual benefit, the SU replies to the PU using the spectral band allocated by the SBS; otherwise the SU keeps silent. Finally, the PU selects the most effective SU as the relaying node, and starts the two-phase cooperative communications.

Physical-Layer Cooperative Communication

To improve the spectrum utilization and communication efficiency by establishing cooperation between one SU and one active PU, we discuss a two-phase cooperative communications scheme by exploiting two-dimensional modulation at the SU to relay the PU's data and transmit its own message orthogonally in the same time slot, as shown in Fig. 1.5. After being selected by the PU as the relaying node, the SU uses QAM to attain orthogonal transmissions [10, 15]. For example, upon receiving the transmitted message from the PU in the first phase, the SU uses in-phase (I) binary phase-shift keying (BPSK) to relay the PU's message with power αP_S and quadrature (Q) BPSK to transmit its own information with power $(1 - \alpha)P_S$ in the second phase as illustrated in Fig. 1.5b, where P_S is the total transmit power of the

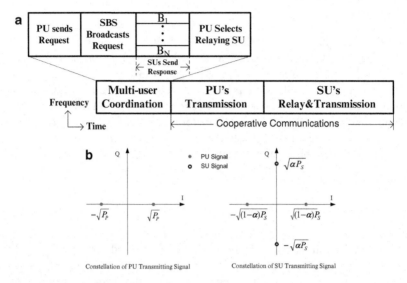

Fig. 1.5 Phase structure and quadrature signaling of cooperative communications between an SU and an active PU. (**a**) Phases of multi-user coordination and cooperative communications. (**b**) Quadrature signaling of the SU relaying and transmission

SU and α is the power allocation factor. Since the I and Q channels are orthogonal to each other, relaying the PU's information and transmitting the SU's own information in the same time slot will not interfere with each other. This enables the SU to cooperatively relay the traffic for the PU while concurrently accessing the same spectrum to transmit the SU's own packets in a two-phase cooperative mode.

Cooperative Spectrum Leasing Between SUs and Inactive PUs

To further improve spectrum efficiency, it is observed that a single temporarily unused spectral band can be leased to multiple SUs even though these SUs are within the same coverage area, while the spectrum is also leased to SUs far away for spectrum reuse purposes. To avoid co-channel interference among SUs, one SU can share the spectrum via cooperative communications with other SUs. One advantage of this scheme is that the leasing cost for each SU decreases since multiple SUs share the leasing price. Another advantage is that SUs can establish collaborative relationships with each other rather than competing as in traditional auction based spectrum leasing. Furthermore, the transmission performance of SUs is improved due to cooperative communications. Therefore, by leasing spectrum to SUs, PUs obtain benefits in terms of monetary reward, and by sharing leasing costs and performing cooperative communications with other SUs, i.e., SUs achieve benefit in terms of accessing spectrum with lower monetary cost and satisfactory transmission performance.

Cooperation Motivation

Since PUs pay to gain the right to use the licensed spectrum, if one PU is to be silent for awhile, it is willing to lease its licensed spectrum during an idle period for financial reward. SUs have the demand to occupy spectrum to fulfill wireless transmissions. This motivates SUs to lease the temporarily unused spectrum from inactive PUs to gain transmission opportunities. In a traditional spectrum leasing framework, one spectrum band can be simultaneously leased to multiple SUs if they are sufficiently far apart from each other as noted in previous sections. By utilizing the quadrature and in-phase channels, one frequency band can be shared by two SUs even when they are within each other's interfering range, e.g., SU_1 and SU_2 as shown in Fig. 1.6b.

Spectrum Leasing

In this section, we discuss a two-tier spectrum leasing scheme, i.e., the SBS leases spectrum from the PBS, and then the SBS leases the spectrum to SUs for secondary transmissions. Specifically, if one PU is to be inactive and decides to lease its licensed spectrum to SUs, the PU sends a leasing supply request to the PBS.[1] The PBS checks whether the PU's spectrum is suitable for leasing. If the PBS agrees with the request of the PU, the PBS sends the leasing supply message to the SBS. After receiving the PBS's leasing information, the SBS decides whether to lease the spectrum band or not.[2] After obtaining the authorization to use the spectrum, the SBS can lease this spectrum to SUs. Specifically, the SBS broadcasts the spectrum information using the spectrum to be leased and allocates orthogonal sub-bands to SUs by dividing the licensed band. Each SU sends bidding and location information to the SBS if an auction is introduced into spectrum leasing. The SBS can lease the spectrum to SUs that are far away, i.e., spectrum spatial reuse, and the SBS can also lease the same spectrum to SUs even when they are within the interfering range. There has been considerable research works on auction based spectrum leasing, and we omit the details of this due to space limitations.

Physical-Layer Cooperative Communications

Assume SU_1 and its neighbor SU_2 win the bidding for leasing one spectral band. We introduce a quadrature signaling based cooperative communications between the SU_1 and SU_2. The idea is similar with the cooperative communications between SUs and active PUs discussed in previous sections. SU_1 and SU_2 both exploit quadrature phase-shift keying (QPSK) modulation to attain orthogonal transmissions, while different channels carry different users' data. For example, SU_1 uses the I channel

---

[1]For a centralized network, the PBS has the spectrum usage information of the PN. The PBS can directly decide to lease temporarily unused spectrum without getting a request from inactive PUs.
[2]The PBS may broadcast the leasing supply message to multiple SBSs for auction, and SBSs then obtain the authority to use spectrum by spectrum bidding.

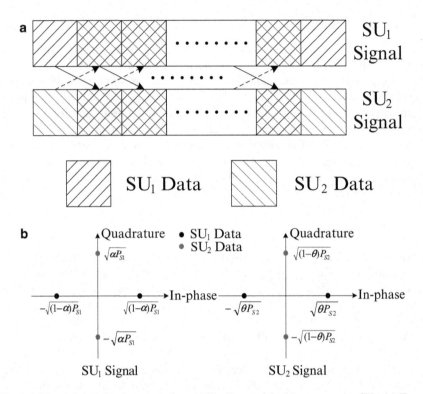

Fig. 1.6 Transmitting signals and quadrature signaling exploited by cooperative SUs. (**a**) Transmitting signals of cooperative SUs. (**b**) Quadrature signaling of cooperative SUs

of QPSK modulation to carry its own data, and SU_2 uses the Q channel of QPSK to carry its own data. As shown in Fig. 1.6, SU_1 and SU_2 start to transmit signals to the SBS simultaneously. After the first data transmission in which SU_1 uses the I channel BPSK and SU_2 uses the Q channel BPSK, SU_1 and SU_2 cooperate to relay data for each other as shown in Fig. 1.6a. In the second data transmission, SU_1 exploits QPSK modulation where the I channel carries its own data with power $(1 - \alpha)P_{S1}$ and the Q channel forwards SU_2's first data with power αP_{S1}, and SU_2 also exploits QPSK modulation, and uses the I channel to forward SU_1's previous data with power θP_{S2} and the Q channel to send its own data with power $(1 - \theta)P_{S2}$ as illustrated in Fig. 1.6b, where P_{S1} and P_{S2} are transmit powers of SU_1 and SU_2, and α and θ are power allocation factors of SU_1 and SU_2, respectively. By using this quadrature signaling, SU_1 and SU_2 can cooperate to communicate with the SBS.

1.4 Conclusions

User cooperation based CRN is a promising framework for improving spectrum efficiency and utilization. In this chapter, we have given a brief overview of cooperative communications between active PUs and SUs, and spectrum leasing between SUs and inactive PUs in CCRN. Following the introduction of these two models, some challenging issues have been presented.

Performing efficient multi-user coordination with low overhead and no interference is significant in the CCRN design. Certain assumptions made in this chapter may be relaxed in future research. For example, although we have limited the coordination process to be a centralized scheme, a distributed one may be more suitable for CCRN.

References

1. I. Akyildiz *et al.* "Next generation/dynamic spectrum access/cognitive radio wireless networks: A survey," *Comput. Netw.*, vol. 50, pp. 2127–2159, May 2006
2. A. Goldsmith, S. A. Jafar, I. Maric, and S. Srinivasa, "Breaking spectrum gridlock with cognitive radios: An information theoretic perspective," *Proc. IEEE*, vol. 97, no. 5, pp. 894–914, May 2009.
3. F. Granelli *et al.*, "Standardization and research in cognitive and dynamic spectrum access networks: IEEE SCC41 efforts and other activities," *IEEE Commun. Mag.*, vol. 48, no. 1, pp. 71–79, Jan. 2010.
4. J. Wang, M. Ghosh, and K. Challapali, "Emerging cognitive radio applications: A survey," *IEEE Commun. Mag.*, vol. 49, no. 3, pp. 74–81, Mar. 2011.
5. B. Wang, and K. Liu, "Advances in cognitive radio networks: A survey," *IEEE J. Sel. Top. Sig. Proc.*, vol. 5, no. 1, pp. 5–23, Feb. 2011.
6. X. Huang, T. Han, and N. Ansari, "On Green-Energy-Powered Cognitive Radio Networks," *IEEE Communications Surveys Tutorials*, vol. 17, no. 2, pp. 827–842, 2015.
7. A. Ahmad, S. Ahmad, M. H. Rehmani, and N. Hassan, "A Survey on Radio Resource Allocation in Cognitive Radio Sensor Networks," *IEEE Communications Surveys Tutorials*, vol. 17, no. 2, pp. 888–917, 2015.
8. O. Simeone, I. Stanojev, S. Savazzi, Y. Bar-Ness, U. Spagnolini, and R. Pickholtz. "Spectrum leasing to cooperating secondary ad hoc networks", *IEEE J. Sel. Areas Commun.*, vol. 26, no. 1, pp. 203–213, Jan. 2008.
9. S. Hua, H. Liu, M. Wu, and S. Panwar, "Exploiting MIMO antennas in cooperative cognitive radio networks," in *Proc. INFOCOM*, Shanghai, Apr. 2011.
10. B. Cao, L. X. Cai, H. Liang, J. Mark, Q. Zhang, H. V. Poor, and W. Zhuang, "Cooperative cognitive radio networking using quadrature signaling," in *Proc. INFOCOM*, Orlando, Apr. 2012.
11. S. Sodagari, A. Attar, and S. Bilén, "On a truthful mechanism for expiring spectrum sharing in cognitive radio networks," *IEEE J. Sel. Areas Commun.*, vol. 29, no. 4, pp. 856–865, Apr. 2011.

12. M. Al-Ayyoub, and H. Gupta, "Truthful spectrum acutions with approximate reveune," *Proc. IEEE INFOCOM*, Shanghai, Apr. 2011.
13. J. Peha, "Sharing spectrum through spectrum policy reform and cognitive radio," *Proc. of IEEE*, vol. 97, no. 4, pp. 708–719, Apr. 2009.
14. K. Lataief, and W. Zhang, "Cooperative communications for cognitive radio networks," *Proc. of IEEE*, vol. 97, no. 5, pp. 878–893, May. 2009.
15. V. Mahinthan, J. Mark, and X. Shen, "A cooperative diversity scheme based on quadrature signaling," *IEEE Trans. on Wirel. Commun.*, vol. 6, no. 1, pp. 41–45, Jan. 2007.

Chapter 2
Orthogonal Signaling Enabled Cooperative Cognitive Radio Networking

Abstract In this chapter, a cross-layer two-phase time division multiple access (TDMA) cooperation framework for primary users (PUs) and secondary user (SUs) in a cooperative cognitive radio network (CCRN) is proposed and analyzed. Specifically, the cooperation framework in which the SU uses the two-dimensional orthogonal modulation for leveraging two degrees of freedom to relay the PU's packet and transmit its own data orthogonally in the same time slot is firstly explored. To evaluate the cooperation performance of the proposed framework, a weighted sum throughput maximization problem is then formulated. With the help of primal-dual sub-gradient algorithms, the optimization problem is solved to obtain closed-form solutions to the optimal powers and allocation of the PU and the SU for both the amplify-and-forward (AF) and decode-and-forward (DF) relaying modes. Cooperative regions based on channel state information are given and discussed, and a cross-layer multi-user coordination for a PU to select a relaying SU for both AF and DF are presented. Extensive simulation results validate the theoretical analysis and show that the proposed two-phase TDMA cooperation framework can achieve mutual benefit in the CCRN.

With the rapid advances in wireless communications and multimedia services, the demand for radio frequency spectrum has been increasing at an explosive rate. To support the evolution of wireless communications, one critical issue to be addressed is how to most efficiently use the available radio spectrum. Traditionally, since the amount of useful radio spectrum is limited, frequency bands are strictly managed and statically allocated in a command-and-control way by regulatory bodies. Within these enforcing policies, only licensed users, referred to as primary users (PUs), can occupy assigned spectral bands. Although the legacy fixed spectrum access (FSA) can effectively avoid or minimize interference among different PUs, as reported in [1], the licensed radio spectrum is significantly underutilized by PUs. The available spectrum in economically feasible frequency regions is already limited, and this spectrum underutilization further exacerbates the situation. In order to address spectral underutilization and inefficiency resulting from FSA, dynamic spectrum access (DSA) methods have been proposed to enable unlicensed users to opportunistically share the licensed spectrum on a non-interference basis.

© The Author(s) 2016

B. Cao et al., *Cooperative Cognitive Radio Networking*, SpringerBriefs in Electrical and Computer Engineering, DOI 10.1007/978-3-319-32881-2_2

Unlicensed users are referred to as secondary users (SUs), also known as cognitive radio users (CRUs). A network consisting of CRUs is referred to as a cognitive radio network (CRN) [2].

In a CRN, by sensing the availability of unused spectral bands, SUs are able to transmit their data in unused licensed bands. Carrying out accurate spectrum sensing is a challenging task; moreover, an SU must abandon its transmission as soon as it detects the presence of PU's transmission in the same spectral band. Unless another idle spectral band is found, the SU has to terminate its transmission entirely. An alternative approach for SUs to gain transmission opportunities using licensed bands is to provide tangible service to PUs to enhance their transmissions, thereby gaining a reward for transmission in licensed bands. A CRN that achieves mutual benefit for both PUs and SUs through cooperation is referred to as a cooperative cognitive radio network (CCRN).

There has been considerable research results on CCRNs [3–22], considering techniques such as pooling detection results from a group of SUs to improve spectrum sensing accuracy [3–5]. Although cooperation can be among SUs or among PUs, from a cognitive radio point of view, the most appropriate cooperation to address spectrum underutilization is cooperation between PUs and SUs to achieve mutual benefits.

There are different ways in which PUs and SUs can cooperate [11–22]. Most of the work reported in the literature use a three-phase strategy in a time-division multiple access (TDMA) mode [12–15]. Specifically, a PU transmits its packets in the first phase, and one or more SUs cooperatively relay the PU's traffic in the second phase. As a reward, the cooperating SUs are allowed to transmit in the third phase. In [22], a two-phase CCRN framework based on multiple-input-multiple-output (MIMO) technology to facilitate transmissions by SUs via multiple antennas is proposed. However, MIMO technology is not widely and readily deployed in CCRNs due to hardware and cost constraints of user devices.

To improve spectrum utilization and efficiency, we propose a two-phase TDMA framework for a PU and an SU to cooperate whereby the SU uses two-dimensional modulation to relay the PU's packet and transmit its own data simultaneously in orthogonal channels. Specifically, the SU uses orthogonal modulation to attain interference-free transmissions in the in-phase and quadrature modulation channels. In this cooperative framework, the PU uses in-phase modulation to transmit its packets, while the SU uses in-phase modulation to relay the PU's packets and quadrature modulation to transmit its own data without interference.

In cooperative communication, the relaying SU can provide a better quality path for the PU than the direct path, or an additional propagation path to facilitate diversity reception at the destination. In the latter case, the receiving PU can appropriately combine the signals from the direct and relayed paths to obtain diversity gain. Relaying is a signal forwarding service, which can be performed using an amplify-and-forward (AF) or a decode-and-forward (DF) relaying mode. Whether the SU can offer feasible relaying service to the PU strictly depends on the quality of the propagation channels between the transmitting PU and the relaying

SU, and between the relaying SU and the receiving PU. Since the PU is licensed to use the dedicated spectrum, the PU has the right to select the most appropriate SU as a relay.

In this chapter, we propose a two-phase TDMA scheme for cooperation between a PU and an SU in which the SU uses orthogonal modulation to relay the PU's signal and to transmit its own data without interfering with each other. As shown in Fig. 2.1, the source PU sends information to a primary base station (PBS)or another PU, and the relaying SU forwards the PU's message and sends its own data to a secondary base station (SBS) or another SU. In this context, the basic configuration is a four-node network. A weighted sum throughput of the cooperating PU and SU is used to evaluate the user cooperation performance of the proposed framework. The objective is to maximize the weighted sum throughput subject to certain performance gain and power constraints for the PU and SU, e.g., the PU should attain a certain throughput gain by cooperating with the chosen SU, which is in turn rewarded with a satisfactory spectral access opportunity, e.g., the SU can achieve a throughput that is larger than its minimum throughput requirement. The main aims of this chapter are four-fold:

- Presentation of an efficient two-phase TDMA cooperation framework: By exploiting the degree of freedom provided by orthogonal modulation, the SU simultaneously relays the PU's message and transmits its own data in the same time slot without mutual interference. Based on the proposed two-phase cooperation framework, we analyze the signal-to-noise ratios (SNRs) in both the primary and secondary links for the AF and DF relaying modes.
- Quantification of the mutual benefit for the PU and SU in terms of achievable throughput: To achieve the mutual benefit, we formulate a weighted sum throughput maximization problem under throughput requirements and power constraints of the PU and SU, e.g., for the PU to attain a given throughput gain and for the SU to get a spectrum access opportunity to achieve a satisfactory throughput such that the total power is bounded at a prescribed level. It is shown that the metric of the weighted sum throughput is general and can represent the throughput of both/either the PU and/or SU by varying the weighting parameter.
- Solution to the optimization problem: With the help of primal-dual sub-gradient algorithms, closed-form solutions to the optimal powers and power allocation of the PU and SU are obtained, and the desirable cooperative regions obtained from channel quality of the associated links for cooperation between the PU and SU under AF and DF are analyzed.
- Selection of the most appropriate relaying SU: By using the cooperative regions, a simple and interference-free multi-user coordination scheme for the PU to select the most appropriate SU amongst multiple SU candidates as the relay is presented.

The remainder of the chapter is organized as follows. Section 2.1 describes the system model and problem formulation. In Sect. 2.2, we solve the optimization problem by obtaining closed-form solutions and deriving the cooperative regions. Based on the derived cooperative regions, we address a cross-layer multi-user coordination scheme for the PU to select the most appropriate SU as the cooperation partner in Sect. 2.3. Extensive simulation results are provided in Sect. 2.4 to demonstrate the effectiveness of the proposed framework. Finally, Sect. 2.5 concludes this chapter.

2.1 System Model and Problem formulation

In this section, the networking architecture, physical layer signaling and cooperation processes are described. The SNR metric of both the primary and secondary links for the AF and DF relaying modes are given. A weighted sum throughput optimization problem subject to throughput requirements and power constraints is formulated.

2.1.1 System Model

We consider a CCRN consisting of one primary network (PN) with a PBS and multiple PUs, and one secondary network (SN) with an SBS and multiple SUs, as shown in Fig. 2.1. In the PN, the PBS allocates network resources to PUs, e.g., frequency bands, time slots, such that PUs can exclusively access assigned

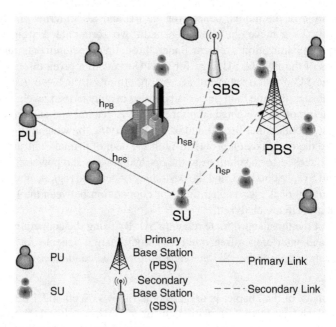

Fig. 2.1 Cooperation between PUs and SUs

Fig. 2.2 Time frame structure for the PU cooperating with one SU

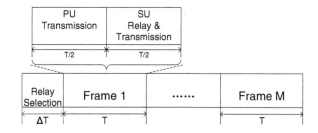

spectral bands or time slots without interfering with each other. When a PU cannot successfully connect with its intended destination such as the PBS or another PU, e.g., if the PU is located far away from the PBS or the direct link between the PU and the PBS is blocked by buildings, the PU has the motivation to select an SU to help relay the primary traffic by yielding a fraction of its spectrum for the SU to use. On the other hand, SUs in the neighborhood of the PU will check their channel conditions with the PU, the PBS, and their own destinations, and decide whether to cooperate with the PU according to the achievable cooperation benefit in terms of throughput and power constraints.

As shown in Fig. 2.2, channel time is divided into frames, each of duration T. M consecutive frames form a superframe. A prefix of duration ΔT for relay selection is appended at the start of each superframe.[1] Each frame is partitioned into two slots (or phases), each of duration $T/2$. The first slot is for PU transmission and the second slot is for the relay to forward the PU's message and to transmit its own data. Relay selection will be described in Sect. 2.3.

In the first phase of each frame, the PU sends its packets to the relay and the destination using in-phase modulation. Upon receiving the PU's packet, in the second phase, the SU forwards the PU's packets using an in-phase channel with power $(1 - \alpha)P_S$ and transmits its own data using a quadrature modulation with power αP_S. The quadrature modulation framework for the PU and the SU are shown in Fig. 2.3, in which Fig. 2.3a is the orthogonal modulator at the relaying SU transmitter to relay for the PU and transmit the SU's own information. Figure 2.3b shows the quadrature demodulation of the SU's own information at the SU receiver, and Fig. 2.3c shows the in-phase demodulation of the PU's data at the PU receiver, respectively. Since these two modulations are orthogonal, there is no mutual interference between the PU's and SU's concurrent transmissions.

In the following subsections, we analyze the received SNRs of the primary and secondary links when the AF and DF relaying modes are used.

---

[1]For convenience, we consider the case in which relay selection is performed regularly at the start of each superframe. In general, relay selection may be carried out as necessary, in which case the prefix would be inserted accordingly.

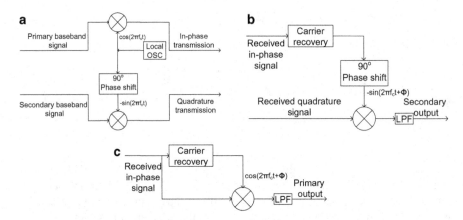

Fig. 2.3 Block diagram of enabling orthogonal modulations in CCRN. (**a**) Quadrature modulator at the relaying SU transmitter. (**b**) Quadrature demodulator at the SU receiver. (**c**) In-phase demodulator at the PU receiver

2.1.2   SNR Analysis of Four-Node Cooperative Communications

Consider a four node cooperative communication scenario in which a PU communicates with the PBS, and an SU communicates with the SBS as shown in Fig. 2.1. The channel between any two users is assumed to follow quasi-static Rayleigh flat fading, i.e., the channel gains are invariant during the time duration $\Delta T + T$. Let h_{PS}, h_{PB}, h_{SP}, and h_{SB} be respectively the channel gains from the PU to SU, PU to PBS, SU to PBS, and SU to SBS.

2.1.2.1   The AF Relaying Mode

Consider a situation in which an SU uses the AF relaying mode and orthogonal modulations to cooperate with a PU. With the orthogonal modulations, information transmission and relaying requires only two phases. In the first phase, the PU transmits an $L-$bit packet x_P using in-phase modulation with power P_P to the SU and PBS, where $E\{x_P^2\} = 1$, and $E\{\bullet\}$ denotes expectation. After receiving the PU's signal, the SU amplifies and forwards it using the in-phase modulator with power $(1 - \alpha)P_S$, and transmits its own data x_S using the quadrature modulator with power αP_S, as shown in Fig. 2.3a.

By receiving the PU's signal, allocating power for relaying the PU's signal and sending the SU's own signal, the transmitted signal of the SU in the second phase is given by

$$y_{S\_AF} = \sqrt{\frac{(1-\alpha)P_S}{|h_{PS}|^2 P_P + \sigma^2}}(h_{PS}\sqrt{P_P}x_P + w) + j\sqrt{\alpha P_S}x_S$$

where w is the additive noise with zero mean and variance σ^2, and j is the imaginary unit denoting the quadrature component.

As shown in Fig. 2.3c, the PBS can extract PU's signal using in-phase modulator. Assuming the maximal ratio combining (MRC) technique is applied by the PBS to decode the PU's packet, the SNR of the received primary signal after combining at the PBS is obtained as

$$SNR_{PB\_AF} = \gamma_{PB}P_P + \frac{\gamma_{SP}\gamma_{PS}P_P P_S (1-\alpha)}{\gamma_{SP}P_S (1-\alpha) + \gamma_{PS}P_P + 1} \tag{2.1}$$

where $\gamma_{ik} = \frac{|h_{ik}|^2}{\sigma^2}$, $i \in \{P, S\}$ and $k \in \{S, P, B\}$, is the channel gain-to-noise ratio (CNR) of the link between node i and node k.

Similarly, the SNR of the received secondary signal at the SBS can be written as $SNR_{SB\_AF} = \alpha\gamma_{SB}P_S$.

2.1.2.2   The DF Relaying Mode

Compared to AF, DF relaying is more complex, but can correct errors before forwarding. Here, in the first phase, the PU broadcasts the signal x_P using in-phase modulation with power P_P. If the relaying SU detects no errors or the detected errors are correctable, the SU relays the PU's packet and transmits its own data in the second phase using the two orthogonal channels provided by orthogonal modulations. Let P_e be the probability that the detected errors are not correctable so that the PU's packet cannot be relayed. Let $P_b = \frac{1}{2}(1 - \sqrt{\frac{P_P\gamma_{PS}}{P_P\gamma_{PS}+1}})$ be the bit error rate (BER) of the PU's signal received by the SU. Assuming that bit errors are independent of each other, we have $P_e = 1 - (1 - P_b)^L \approx LP_b$ where the approximation is for small P_b and large L. In the context of using DF relaying, the transmitted signal of the SU in the second phase is given by

$$y_{S\_DF} = \begin{cases} \sqrt{(1-\alpha)P_S}x_P + j\sqrt{\alpha P_S}x_S, & \text{w.p. } 1 - P_e \\ 0, & \text{w.p. } P_e. \end{cases}$$

Therefore, the SNRs received at the PBS and SBS are given by

$$SNR_{PB\_DF} = \begin{cases} P_P\gamma_{PB} + (1-\alpha)P_S\gamma_{SP}, & \text{w.p. } 1 - P_e \\ P_P\gamma_{PB}, & \text{w.p. } P_e \end{cases} \tag{2.2}$$

and

$$SNR_{SB\_DF} = \begin{cases} \alpha P_S \gamma_{SB}, & \text{w.p. } 1 - P_e; \\ 0, & \text{w.p. } P_e \end{cases} \tag{2.3}$$

respectively, where 0 implies that no useful signal is received by the SBS, i.e., the SU does not transmit.

2.1.3   Problem Formulation

In this subsection, we formulate a throughput optimization problem under certain constraints in a CCRN. Instead of maximizing either the PU's throughput, or the SU's throughput, we propose a general metric of weighted sum throughput, namely C_W which incorporates both the PU's and SU's throughputs to evaluate the cooperation performance. The optimization problem is formulated as

$$\max_{\alpha, P_P, P_S} C_W = (1 - \zeta)C_{PB} + \zeta C_{SB} \tag{2.4}$$

Subject to :

$$C_{PB} \geq KC_{Pd} \geq C_{PT}, \quad \text{for } K \geq 1$$

$$C_{SB} \geq C_{ST}$$

$$P_P + P_S \leq P_M$$

$$0 < \alpha < 1$$

where ζ is a weighting parameter that strikes a balance between the PU's and SU's throughputs. In general, the parameter ζ reflects the cooperation needs of the PU. For example, if a PU is experiencing a deep fade and cannot communicate with the PBS without the help of SUs, a larger ζ value may be used to favor SUs, and vice versa. In the extreme cases when $\zeta = 0$ or $\zeta = 1$, the objective function is simplified to maximize the PU's or SU's throughput, as in [12, 16, 18], and [22]. Let $C_{Pd} = \log_2(1 + P_P\gamma_{PB})$ be the achievable throughput of the PU to the PBS without cooperating with an SU, and K be the throughput gain that the PU wants to achieve through cooperation. The PU needs to cooperate with SUs if the achievable C_{Pd} is below its minimum throughput requirement, C_{PT}. The first constraint in (2.4) shows that, by cooperating with an SU, the PU can achieve a throughput greater than the threshold C_{PT}. The second constraint represents that the achievable throughout of the SU should meet its minimum throughput requirement, C_{ST}. Otherwise, the SU may not be able to establish a connection with the SBS so that the secondary transmission becomes useless. These two constraints represent the cooperation benefit that motivates the PU and SU to cooperate with each other. The third constraint indicates that the total transmission power of the PU and SU should be bounded by P_M. Lastly, the parameter α, the fraction of the power that an

SU uses for transmitting its own traffic, should be selected from (0,1). In a special case when $\alpha = 1$, the SU transmits only its own data without relaying the PU's packet; alternatively, when $\alpha = 0$, the SU only helps forward the PU's packet without transmitting its own data, which is a conventional relaying scenario.

2.2 Optimization and Analysis of the Proposed Framework

In this section, we solve the constrained optimization problem in (2.4) by applying the Karush-Kuhn-Tucker (KKT) conditions. Firstly, the optimal power levels of the PU and SU, denoted respectively by P_P^* and P_S^*, are obtained by considering the throughput and power constraints. Given P_P and P_S, we derive the optimal power allocation parameter α^* to simplify the computational complexity. In addition, based on the channel conditions, we give the cooperative conditions of the PU and SU for the AF and DF relaying modes.

2.2.1 Optimization of the AF Relaying Mode

In a fading channel with additive Gaussian noise, the achievable throughput of the PU and the SU are respectively given by $C_{PB\_AF} = B\log_2(1 + SNR_{PB\_AF})$ and $C_{SB\_AF} = B\log_2(1 + SNR_{SB\_AF})$, where B is a system parameter determined by the available bandwidth and the transmission efficiency. For example, if the time duration of each phase is $\frac{T}{2}$ in the proposed framework, then we have $B = \frac{1}{2}$.

Suppose the PU wants to cooperate with an SU if the achievable throughput by cooperation is higher than that of the direct link, i.e., we have $K = 1$ in (2.4). By using the Lagrange multiplier method, we can obtain closed-form optimal expressions for values of P_P^*, P_S^*, and α^*; the detailed derivations are given in Appendix.

Note that the computational complexity involved in obtaining the optimal solution is high. To simplify the analysis, we assume the transmission power values of the PU and SU are given, which is acceptable in most practical communication systems without adaptive power control. In this context, we need to determine only the optimal power allocation for the SU to relay the PU's packet and to transmit its own data, i.e., the optimal α. To this end, we check the first two constraints in (2.4). By mathematical manipulation, the following inequalities are obtained:

$$\gamma_{SP} > \frac{\gamma_{PB}(\gamma_{PS}P_P + 1)(1 + \gamma_{PB}P_P)}{P_S(\gamma_{PS} - \gamma_{PB}^2 P_P - \gamma_{PB})} \tag{2.5}$$

$$\gamma_{PS} > \gamma_{PB}^2 P_P + \gamma_{PB} \tag{2.6}$$

$$\gamma_{SB} > \frac{\exp(2C_{ST}\ln 2) - 1}{P_S}. \tag{2.7}$$

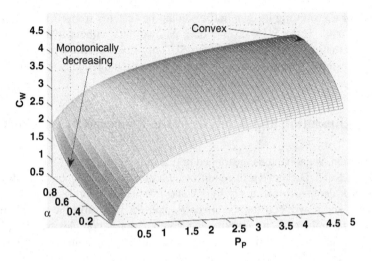

Fig. 2.4 C_W for different α and P_P

Equations (2.5)–(2.7) represent the cooperative conditions for the PU and SU when the AF mode is used. That is, if the quality of channels, including the channel between the SU and PBS, between the SU and SBS, and between the PU and SU, are above the thresholds derived in (2.5)–(2.7), the PU and SU can cooperate with each other and both can achieve certain cooperation benefit that fulfill their throughput requirements under power constraints. In general, the SU checks the cooperative conditions based on the estimated channel state information (CSI).

If the SU is within the cooperative conditions when all the channel states are above the corresponding thresholds, the SU calculates the optimal α by calculating the maximum C_W for $\alpha \in (0, 1)$. As shown in Fig. 2.4, C_W decreases with respect to (w.r.t.) α when P_P is small, and the maximum C_W is achieved when $P_P = 5$ mW.

When C_W has one extreme point w.r.t. $\alpha \in (0, 1)$ under certain P_P, by solving $\frac{\partial C_W}{\partial \alpha} = 0$, α_{AF}^* is obtained by

$$\alpha_{AF}^* = \frac{R(\gamma_{PS}P_S + P_P\gamma_{PS} + 1)}{F\gamma_{PS}P_S} \tag{2.8}$$

where

$$R = \zeta\gamma_{PS}P_S + \zeta P_P\gamma_{PB}\gamma_{PS}P_S + \zeta P_P P_S\gamma_{PS}\gamma_{SB}$$
$$+ 2\zeta P_P\gamma_{PS} - P_P\gamma_{PS} + \zeta P_P^2\gamma_{PB}\gamma_{PS} + \zeta + \zeta P_P\gamma_{PB}$$

and

$$
\begin{aligned}
F = & \zeta \gamma_{PS} P_S + \zeta P_P \gamma_{PB} \gamma_{PS} P_S + \zeta P_P P_S \gamma_{PS} \gamma_{SB} \\
& - P_P \gamma_{PS} + \zeta P_P \gamma_{PB} + \gamma_{PS} P_P^2 \gamma_{SP} + 2\zeta P_P \gamma_{PS} \\
& + \zeta P_P^2 \gamma_{PB} \gamma_{PS} + \zeta + P_P \gamma_{SP}.
\end{aligned}
$$

Since C_{PB_AF} decreases w.r.t. $\alpha \in (0, 1)$, and C_{SB_AF} increases w.r.t. $\alpha \in (0, 1)$, if C_W is an increasing function w.r.t. $\alpha \in (0, 1)$, in order to improve the opportunity of being selected by the PU, the SU should provide the throughput gain to the PU as high as the SU can. By doing so, the SU selects an α value that meets its own throughput requirement constraint as α_{AF}^* which is derived as

$$
\alpha_{AF}^* = \frac{(\gamma_{SB} P_S + P_P \gamma_{PS} + 1)(e^{2C_{ST} \ln 2} - 1)}{\gamma_{SB} P_S (e^{2C_{ST} \ln 2} + P_P \gamma_{PS})}. \tag{2.9}
$$

2.2.2 Optimization of the DF Relaying Mode

When the DF mode is applied, with probability P_e the SU does not relay the PU's packet when the detected errors are not correctable, so the achievable throughput of the PU is determined by the direct transmission from the PU to the PBS. In the case when the detected errors are correctable the achievable throughput is a combination of the direct and the relayed paths. Therefore, the throughput of the PU is given by

$$
C_{PB\_DF} = \begin{cases} B \log_2[1 + P_P \gamma_{PB} + (1 - \alpha) P_S \gamma_{SP}], & \text{w.p. } 1 - P_e \\ B \log_2(1 + P_P \gamma_{PB}), & \text{w.p. } P_e. \end{cases}
$$

Similarly, the throughput of the SU in the second phase of cooperative transmission is

$$
C_{SB\_DF} = \begin{cases} B \log_2(1 + \alpha P_S \gamma_{SB}), & \text{w.p. } 1 - P_e \\ 0, & \text{w.p. } P_e. \end{cases} \tag{2.10}
$$

The expected throughput of the PU and SU can be derived as

$$
\begin{aligned}
\hat{C}_{PB\_DF} = & B(1 - P_e) \log_2[1 + P_P \gamma_{PB} + (1 - \alpha) P_S \gamma_{SP}] \\
& + B P_e \log_2(1 + P_P \gamma_{PB})
\end{aligned} \tag{2.11}
$$

and

$$
\hat{C}_{SB\_DF} = B(1 - P_e) \log_2(1 + \alpha P_S \gamma_{SB}). \tag{2.12}
$$

Similar to that in the AF case, the values of P_P^*, P_S^*, and α^* can be jointly obtained by applying the KKT conditions to solve the optimization problem. We can simplify the analysis by deriving α^* for the given P_P and P_S. The Lagrange function is given by

$$\mathcal{L}(\alpha) = (1 - \zeta)\hat{C}_{PB\_DF} + \zeta\hat{C}_{SB\_DF} + \lambda_2[C_{ST} - \hat{C}_{SB\_DF}] \\ + \lambda_1[C_{PB} - \hat{C}_{PB\_DF}]. \tag{2.13}$$

By setting $B = \frac{1}{2}$ and solving the Lagrange function, the optimal α can be obtained as

$$\alpha_{DF}^* = \frac{A}{\gamma_{SP}\gamma_{SB}P_S(\lambda_1 + \lambda_2 - 2)} \tag{2.14}$$

where

$$A = \gamma_{SP}(2 - 2\zeta - 2\zeta\gamma_{SB}P_S - \lambda_1 + \lambda_2\gamma_{SB}P_S) \\ + \gamma_{SB}(\lambda_2 + \lambda_2 P_P\gamma_{PB} - 2\zeta - 2\zeta P_P\gamma_{PB})$$

and λ_1 and λ_2 can be iteratively obtained as

$$\lambda_1^{(n+1)} = [\lambda_1^{(n)} + \mu^{(n)}(C_{PB} - \hat{C}_{PB\_DF}^{(n)})]^+ \tag{2.15a}$$

$$\lambda_2^{(n+1)} = [\lambda_2^{(n)} + \mu^{(n)}(C_{ST} - \hat{C}_{SB\_DF}^n)]^+. \tag{2.15b}$$

Similarly, for given P_P and P_S, the cooperative conditions are derived from the throughput constraints of the PU and SU, which are given by

$$\gamma_{SP} > \frac{P_P\gamma_{PB}(1 + P_P\gamma_{PB})}{P_S} \tag{2.16}$$

$$\gamma_{SB} > (e^{(\ln 2)C_{ST}\frac{\frac{1}{\hat{\gamma}_{PS}}}{\sqrt{\frac{\hat{\gamma}_{PS}}{\hat{\gamma}_{PS}+1}}}} - 1)P_S^{-1} \tag{2.17}$$

where $\hat{\gamma}_{PS} = \frac{L^2 P_P\gamma_{PS}}{(L^2-1)P_P\gamma_{PS}+1}$.

Notice that there are only two thresholds for the DF relaying mode, while there are three thresholds for the AF case. When DF is applied, the SU does not need to check the cooperative conditions if the transmission from the PU to the SU fails. On the contrary, given that the SU successfully receives the packet from the PU, the SU needs only to check whether its transmitting and forwarding channels, i.e., channels between the SU and SBS, and between the SU and PBS, are sufficiently good to satisfy the throughput requirements of both cooperators. Therefore, the SU needs only to check the channel thresholds of γ_{SP} and γ_{SB} in the DF case.

If the SU is within the cooperative conditions, i.e., both γ_{SP} and γ_{SB} are above the thresholds, then the SU calculates the optimal α by calculating the maximum C_W for $\alpha \in (0, 1)$. The optimal α value can be obtained as

$$\alpha_{DF}^* = \frac{(\zeta - 1)\gamma_{SP} + \zeta\gamma_{SB}(1 + P_P\gamma_{PB} + P_S\gamma_{SP})}{\gamma_{SB}P_S\gamma_{SP}} \tag{2.18}$$

if C_W has one extreme point when $\alpha \in (0, 1)$, or

$$\alpha_{DF}^* = \left(e^{C_{ST}(\ln 2)\frac{1}{\sqrt{\frac{\gamma_{PS}}{\gamma_{PS}+1}}}} - 1 \right) \gamma_{SB}^{-1}P_S^{-1} \tag{2.19}$$

if C_W is an increasing function of $\alpha \in (0, 1)$.

2.3 Multi-User Coordination

Based on the derived cooperative conditions of the PU and SU, we propose a cross-layer multi-user coordination scheme for the PU to select an appropriate SU as the relay. In the SN consisting of N SUs, when the PBS detects that a PU is suffering from severe channel fading during its communication with the PBS, the PU should select the most effective SU among all SU candidates as its relaying partner.

In general, in the uplink, if the PBS detects that the outage probability of the direct link from the PU to the PBS is over the minimum outage probability requirement to support the PU's service, or the communication efficiency is not high enough to meet the minimum flow requirement, the PBS informs the PU to find a helpful partner to improve its transmission performance in terms of reliability or efficiency.

In particular, if the PBS detects that the primary link is of low quality, then the PBS broadcasts a notice to the PU and SN to notify that the current PU needs to find a relaying SU as a partner. After getting this notice, the SBS will prepare to allocate resources to SUs for multi-user coordination. Meanwhile, by hearing this notice message from the PBS, all SU candidates can estimate the channel states between the PBS and themselves. For the PU, after receiving the cooperation notice message from the PBS, the PU plans to perform multi-user selection with the help of the SBS.

In this context, when the PU receives the notice from the PBS and checks that its current transmission rate C_{Pd} is indeed lower than the minimum throughput requirement, i.e., $C_{Pd} < C_{PT}$, the PU will send a cooperation request message to the SBS. Since the PU is to establish cooperation with the SN, the PU can yield its licensed spectrum to the SN for the multi-user coordination and wait for responses from potential SU candidates. Therefore, the multi-user coordination will not interfere with other PUs, since each PU owns orthogonal frequency bands or time slots as described above. SUs can also receive the PU's cooperation request if the channel states between them are relatively good. In this way, SUs can estimate the channel gains between the PU and themselves.

The cooperation request message includes the PU's expected throughput requirement, and the allocated time of using the licensed frequency bands, i.e., the length of the cooperation phase. Upon receiving the request message from the PU, the SBS multi-casts the PU's request to N SUs by using N orthogonal channels divided from the licensed frequency band dedicated to the PU. After receiving the request, each SU first checks whether the cooperation is mutually beneficial for both the PU and itself. Specifically, to ensure that both the PU's and the SU's throughput requirements can be satisfied, each SU checks the cooperative conditions as derived in (2.5)–(2.7) and (2.16)–(2.17) for the AF and DF modes, respectively. As discussed in Sect. 2.2, the AF and DF modes result in different cooperative regions, and the SU always selects a relaying mode that achieves a greater throughput gain for the PU. An eligible relay candidate SU then computes the optimal power ratio α by using (2.8) or (2.9) for AF, or (2.18) or (2.19) for DF. Only potential candidate SUs satisfying the cooperative conditions respond to the PU, using the orthogonal sub-channels allocated by the SBS. The response includes the achievable throughput of the PU by cooperating with the SU and the selected cooperation mode, i.e., AF or DF. Based on the collected response messages from multiple SUs, the PU selects the SU that can provide the greatest throughput gain as its cooperator, i.e., the PU selects the SU_{l*}, with

$$l^* = \arg \max_{l^* \in \{l_1, l_2\}} \{C_{PB\_AF}^{(l_1)}, C_{PB\_DF}^{(l_2)}\} \tag{2.20}$$

where the index l_1 means that SU_{l_1} can provide the greatest throughput gain among SUs using the AF mode, and index l_2 means that SU_{l_2} can provide the greatest throughput gain among SUs using the DF mode. Then the PU and SU initiate a two-phase cooperation, as shown in Fig. 2.2. In the case in which no SU satisfies the cooperation conditions, the PU can only use the direct link to the PBS for transmission.

It can be seen that the signaling overhead for the proposed multi-user coordination includes three parts. The first part is the cooperation triggering process by the PBS and PU, and other relay-selection protocols such as in [12–15] and [22] should also take into account this process. In this part, the PU needs to send the SN the information including current direct link channel gain, transmission power, range of frequency bands, duration of time slots, synchronization information, and expected throughput gain. The communication overhead for this part is fixed. The second part is the optimization process operated by the SN. Specifically, the SBS divides the PU's frequency band into N orthogonal segments and allocates them to all SUs. Potential SU candidates need to estimate the associated channel gains to mechanize the optimal transmission and relaying strategy. CSI estimation should also be performed [12–15, 22]. The last part is for the PU to select the most effective SU as the partner. In this part, the PU needs to collect responses from SU

Fig. 2.5 Throughput
comparison w/wo
cooperation (the AF relaying
mode, $\zeta = 0.3$)

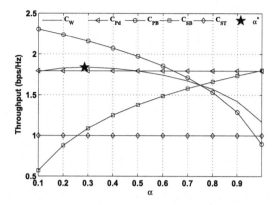

candidates and select the SU that can provide the largest primary throughput. The
communication and information processing overhead is related to the number of
SU candidates. While this process can be operated by the SBS, i.e., after collecting
and comparing all responses from SU candidates, the SBS sends the final selection
result to the PU. In this case, the communication overhead for the PN can be kept at
an accepted level.

2.4   Numerical Results

In this section, the performance of the proposed framework is evaluated via
extensive simulations with Matlab.

Consider that the PU needs to find an SU as a partner to relay the primary
transmission when the quality of the direct link to the PBS is poor. There are four
SU candidates managed by an SBS in the SN. The channel model incorporates both
large-scale attenuation with a path loss exponent from 2 to 7 and small-scale flat
Rayleigh fading. To show the cooperation benefit, the link from the PU to the PBS
suffers from a deep fade defined in the simulation parameters. For simplicity, we
assume both the PU and SU use constant transmission power in the simulations.

To validate the analytical results and the effectiveness of the proposed framework, analytical and simulation results on the optimal α, and the throughput
performance of the PU and SU when using the AF mode are shown in Fig. 2.5.
The CNR values used are $\gamma_{PB} = 2.4679$, $\gamma_{PS} = 3.4421$, $\gamma_{SB} = 6.6273$, and
$\gamma_{SP} = 13.4497$. Due to the severe deep Rayleigh channel fading, the achievable
direct link throughput of the PU is $C_{Pd} = 1.8$ bps/Hz. The powers of the PU and
SU are $P_P = 1$ mW and $P_S = 2$ mW, and the weighting parameter is set as $\zeta = 0.3$.
The SU is assumed to be able to obtain perfect CSI. After getting the CSI, the SU
compares it with the associated cooperative conditions. In this simulation, the CSI
values are within the cooperative conditions when AF is used, and the SU aims to
get the optimal power allocation factor for relaying the PU's traffic and transmitting
its own data.

Fig. 2.6 Throughput comparison w/wo cooperation (the DF relaying mode, $\zeta = 0.3$)

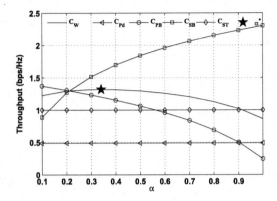

It is shown in Fig. 2.5 that the maximum throughput is achieved when α is around 0.29 in this simulation, and by calculating the optimal according to (2.8) we obtain $\alpha^* \approx 0.286$. In this context, the throughput of the PU and SU are $C_{SB\_AF} \approx 1.1$ bps/Hz and $C_{PB\_AF} \approx 2.2$ bps/Hz, respectively. Inspection shows that the analytical results closely approximate the simulation ones. It is also observed that by cooperation, the PU can achieve 1.22 times the throughput compared with direct transmission while the SU can also achieve a higher throughput than its minimum throughput requirement of $C_{ST} = 1$ bps/Hz.

To show the difference in cooperative conditions of the AF and DF relaying modes, this simulation illustrates how to adapt the relaying mode according to instantaneous CSI. When the CNRs are $\gamma_{PB} = 0.4024$, $\gamma_{PS} = 9.9048$, $\gamma_{SB} = 12.6694$, $\gamma_{SP} = 3.0870$, and power levels of the PU and SU are $P_P = 1$ mW and $P_S = 1$ mW, the AF mode cannot provide a feasible solution for user cooperation, and the SU then checks whether the DF mode can be applied. Figure 2.6 plots the throughput performance of the PU and SU for the DF mode. It can be seen in the figure that the optimal $\alpha^* \approx 0.341$ can be obtained by using (2.18) in this case, while from the simulation it can be seen that the optimal α is around 0.34. In this context, the SU achieves $C_{SB\_DF} \approx 1.6 > C_{ST} = 1$ bps/Hz and the PU achieves $C_{PB\_DF} \approx 1.25 > C_{Pd} = 0.4879$ bps/Hz.

To show the multi-user coordination process, this simulation demonstrates how each SU determines the most suitable relaying strategy after getting the instantaneous CSI values and cooperative conditions for both the AF and DF modes. Consider for example SU_3 among the four SU candidates. Figure 2.7a shows the weighted sum throughput with different relaying modes for SU_3. It is shown that the AF relaying mode achieves a higher weighted sum throughput than DF for $\alpha < 0.86$. When $\alpha^*_{AF} = 0.49$, the achievable throughput of the PU and SU using AF are $C_{PB\_AF3} = 1.29$ and $C_{SB\_AF3} = 1.21$, respectively. For the DF case, $\alpha^*_{AF} = 0.59$, and the achievable throughput of the PU and SU are $C_{PB\_DF3} = 1.17$ and $C_{SB\_DF3} = 1.1$, respectively. As shown in Fig. 2.7b, SU_3 selects the AF relaying mode for cooperation because the PU can achieve a higher throughput gain if SU_3

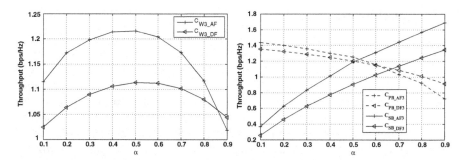

Fig. 2.7 Throughput performance of the AF and DF relaying modes for SU_3 cooperating with the PU. (**a**) C_W for SU_3 using AF and DF relaying modes. (**b**) Primary and secondary throughput for SU_3 using the AF and DF relaying modes

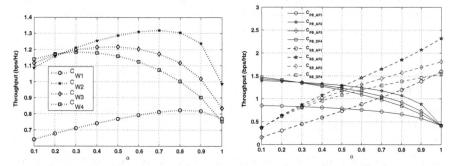

Fig. 2.8 The multi-user coordination process. (**a**) C_W for SU_i ($i = 1, \ldots, 4$) cooperating with the PU. (**b**) Primary and secondary throughput for SU_i ($i = 1, \ldots, 4$) cooperating with the PU

uses AF. Thus, by selecting a relaying mode which can provide a higher throughput gain for the PU, the SU is more likely to be selected as the PU's cooperator and fulfill its own transmission requirement.

After calculating the optimal parameters and the associated throughput, the SU candidates respond to the PU. The throughput of the PU cooperating with different SUs are compared in Fig. 2.8. Due to the poor channel quality, SU_1 does not satisfy its cooperative conditions for either the AF or DF relaying modes. In this case, SU_1 does not respond to the PU. Other SUs, i.e., SU_k ($k = 2, 3, 4$) have relatively good channel quality, and they can send response to inform the PU how much throughout gain the PU can expect to achieve through cooperation, i.e., $C_{PB\_AF2} = 1.23$, $C_{PB\_AF3} = 1.21$ and $C_{PB\_DF4} = 1.42$, and the selected relaying modes, i.e., AF for SU_2 and SU_3, and DF for SU_4. As shown in Fig. 2.8a, SU_2 achieves the maximum C_W among the three SU candidates. However, from Fig. 2.8b, it can be seen that SU_4 can provide the maximum throughput for the PU, and thus the PU cooperates with SU_4 even though the achievable C_W of SU_4 is lower than that of SU_2. With the proposed scheme, it is guaranteed that both the PU and SU can achieve mutual benefit by cooperating with each other.

2.5   Conclusion

We have presented a cooperation framework between a PU and an SU in a cognitive radio network whereby the relaying SU uses orthogonal modulation to forward the PU's signal using the in-phase channel and transmit its own data using the quadrature channel without interfering with each other. In this way, the cooperation framework uses two phases in which the first phase is used by the PU to transmit its information, while the second phase is used by the SU to relay the PU's signal as well as transmit its own data. A weighted sum throughput criterion has been used to optimize network throughput subject to certain power constraints for the PU and SU. The optimal power settings and allocations have been obtained and closed-form solutions have been derived for both the AF and DF relaying modes. The conditions for desirable cooperation for the PU and SU, referred to as cooperative conditions, have been derived, based on which a multi-user coordination scheme has been proposed.

Appendix

Derivation of Optimal Solutions for AF Cooperation

Using the Lagrange multiplier method, the objective function in (2.4) can be expressed as

$$
\begin{aligned}
\mathscr{L}(\alpha, P_P, P_S) = {} & (1 - \zeta)C_{PB\_AF} + \zeta C_{SB\_AF} \\
& + \lambda_1(C_{Pd} - C_{PB\_AF}) + \lambda_2(C_{ST} - C_{SB\_AF}) \\
& + \lambda_3(P_P + P_S - P_M) + \lambda_4(\alpha - 1) - \lambda_5\alpha
\end{aligned}
\tag{2.21}
$$

where $\lambda_i \geq 0$ $(i = 1, .., 5)$ are Lagrange multipliers.

By applying the KKT conditions, a set of equations can be obtained as

$$
P_P > 0, P_S > 0, \text{ and } \quad \lambda_i \geq 0, \text{ for } \quad i = 1, \ldots 5
\tag{2.22a}
$$

$$
\lambda_1[C_{Pd} - \frac{1}{2}\log_2(1 + SNR_{PB\_AF})] = 0
\tag{2.22b}
$$

$$
\lambda_2[C_{ST} - \frac{1}{2}\log_2(1 + SNR_{SB\_AF})] = 0
\tag{2.22c}
$$

$$
\lambda_3(P_P + P_S - P_M) = 0
\tag{2.22d}
$$

$$
\lambda_4(\alpha - 1) = 0
\tag{2.22e}
$$

$$
\lambda_5\alpha = 0
\tag{2.22f}
$$

$$\frac{\partial \mathscr{L}}{\partial \alpha} = 0, \frac{\partial \mathscr{L}}{\partial P_P} = 0 \quad \text{and} \quad \frac{\partial \mathscr{L}}{\partial P_S} = 0. \tag{2.22g}$$

Solving the equation set yields

$$\alpha^* = \frac{M(P_P \gamma_{PS} + \gamma_{SB} P_S + 1)}{N \gamma_{SP} P_S} \tag{2.23}$$

where M and N are given by

$$
\begin{aligned}
M = &(\lambda_4 \zeta - \lambda_1) P_P{}^2 \gamma_{PB} \gamma_{PS} + (2\zeta - \lambda_4) P_P \gamma_{PS} \\
&-(\lambda_1 + P_S \gamma_{SP}) P_P \gamma_{PS} + (\zeta - \lambda_1)(P_P \gamma_{PB} + 1) \\
&+(\zeta \lambda_5 \gamma_{PS} \gamma_{SB} - \lambda_1 \gamma_{PB} \gamma_{SP}) P_S P_P \\
&+(\zeta P_P \gamma_{PB} + \zeta) P_S \gamma_{SB} - \lambda_1 \gamma_{SP} P_S \\
N = &(\zeta \lambda_4 \gamma_{SP} - \lambda_1 \gamma_{SB}) P_P \gamma_{PS} P_S + (\zeta - \lambda_1) P_P \gamma_{PB} \gamma_{SB} P_S \\
&-\lambda_1 \gamma_{SP} P_S + \gamma_{SB} P_S + (\lambda_5 \zeta - \lambda_1) P_P \gamma_{PB} \\
&+(\zeta - \lambda_1) P_P{}^2 \gamma_{PB} \gamma_{PS} - \lambda_1 + 2(\zeta - \lambda_1) P_P \gamma_{PS} \\
&+(1 - \lambda_1) P_P{}^2 \gamma_{PS}{}^2 + \zeta.
\end{aligned}
$$

The quantity P_P^* is the root of the quadratic equation

$$a_4 P_P^4 + a_3 P_P^3 + a_2 P_P^2 + a_1 P_P + a_0 = 0 \tag{2.24}$$

where the coefficients of the equation are

$$
\begin{aligned}
a_4 = &\ln 2 \gamma_{PB} \gamma_{PS}{}^2 \lambda_2 \\
a_3 = &2 \ln 2 \gamma_{PB} \gamma_{PS} \lambda_2 - \lambda_1 \gamma_{PS}{}^2 \gamma_{PB} + \lambda_3 \gamma_{PS}{}^2 \gamma_{PB} \\
&+2 \ln 2 \gamma_{PS}{}^2 \lambda_2 + 2 \ln 2 P_S \gamma_{PB} \gamma_{PS} \gamma_{SP} \lambda_2 \\
a_2 = &(\gamma_{SB} P_S + 1)(\ln 2 P_S \gamma_{PS} \gamma_{SB} \lambda_2 + \ln 2 P_S \gamma_{PB} \gamma_{SP} \lambda_2 \\
&+3 \lambda_3 \ln 2 \gamma_{PS} \lambda_2 + 2 \gamma_{PB} \gamma_{PS} - \lambda_1 \gamma_{PB} \gamma_{PS}) \\
a_1 = &(\gamma_{SB} P_S + 1)(-\gamma_{PS} \zeta \gamma_{SP} P_S - \zeta \gamma_{SP} P_S + \gamma_{PB} \gamma_{SB} P_S \\
&+ \ln 2 P_S \gamma_{SB} \lambda_2 - \zeta \gamma_{PB} + \lambda_2 \ln 2 - 2 \zeta \gamma_{PS} + \lambda_3 \gamma_{PB}) \\
a_0 = &-\lambda_3 \zeta + 2 \lambda_1 \gamma_{SP} P_S - 2 \lambda_3 \zeta \gamma_{SB} P_S \\
&-\zeta \gamma_{SB}{}^2 P_S{}^2 + \lambda_1 \gamma_{SP}{}^2 P_S{}^2 + \lambda_1.
\end{aligned}
$$

Also,

$$P_S^* = \left[\frac{\sqrt{E} - D}{2\lambda_2\lambda_3 \ln 2(\gamma_{PS} + 1)\gamma_{SB}}, 0 \right]^+ \tag{2.25}$$

where

$$D = \lambda_2 \ln 2(2\lambda_3 + 3\gamma_{PS} + 2\lambda_3\gamma_{PB}\gamma_{PS} + \gamma_{PB})$$

$$E = \ln 2\lambda_2(\gamma_{PB} - \lambda_3\gamma_{PS})(\ln 2\gamma_{PB}\lambda_2 - \lambda_3 \ln 2\gamma_{PS}\lambda_2$$

$$+ 4\lambda_3\gamma_{PS}\gamma_{SB} - 4\gamma_{SB}\lambda_1\gamma_{PS} + 4\gamma_{SB} - 4\lambda_1\gamma_{SP}).$$

This quadratic equation can be solved analytically or numerically, and according to the power constraint of the PU, the optimal P_P can be obtained.

Using the Lagrangian relaxation iterative algorithm, we can further iteratively compute λ_i by

$$\lambda_1^{(n+1)} = [\lambda_1^{(n)} + \mu^{(n)}(C_{PB} - C_{PB\_AF}^{(n)})]^+ \tag{2.26}$$

$$\lambda_2^{(n+1)} = [\lambda_2^{(n)} + \mu^{(n)}(C_{ST} - C_{SB\_AF}^{(n)})]^+ \tag{2.27}$$

$$\lambda_3^{(n+1)} = [\lambda_3^{(n)} + \mu^{(n)}(P_P^{(n)} + P_S^{(n)} - P_M)]^+, \tag{2.28}$$

$$\lambda_4^{(n+1)} = [\lambda_4^{(n)} + \mu^{(n)}(\alpha^{(n)} - 1)]^+ \tag{2.29}$$

$$\lambda_5^{(n+1)} = [\lambda_5^{(n)} + \mu^{(n)}\alpha^{(n)}]^+ \tag{2.30}$$

where $[x]^+ = \max(x, 0)$, n is the iteration index and $\mu^{(n)}$ is a sequence of scalar step sizes. Substituting (2.26)–(2.30) into (2.23)–(2.25), we can obtain α^*, P_P^*, and P_S^*.

References

1. FCC, "Spectrum policy task force," *ET Docket 02-135*, Nov. 2002.
2. S. Haykin, "Cognitive radio: Brain-empowered wireless communications," *IEEE J. Sel. Areas Commun.*, vol. 23, no. 2, pp. 201–220, Feb. 2005.
3. S. Mishra, A. Sahai, and R. Brodersen, "Cooperative sensing among cognitive radios," in *Proc. IEEE Int'l Conf. Commun.*, Istanbul, Turkey Jun. 2006.
4. G. Ganesan, and Y. Li, "Cooperative spectrum sensing in cognitive radio, part I: Two user networks," *IEEE Trans. Wirel. Commun.*, vol. 6, no. 6, pp. 2204–2213, Jun. 2007.
5. K. Ben Lataief, and W. Zhang, "Cooperative communications for cognitive radio networks," *Proc. IEEE*, vol. 97, no. 5, pp. 878–893, May 2009.
6. J. Jia, J. Zhang, and Q. Zhang, "Cooperative relay for cognitive radio networks," in *Proc. INFOCOM*, pp. 2304–2312, Rio de Janeiro, Brazil, Apr. 2009.
7. Q. Zhang, J. Jia, and J. Zhang, "Cooperative relay to improve diversity in cognitive radio networks," *IEEE Commun. Magazine*, vol. 47, no. 2, pp. 111–117, Feb. 2009.

8. L. Li, X. Zhou, H. Xu, G. Li, D. Wang, and A. Soong, "Simplified relay selection and power allocation in cooperative cognitive radio systems," *IEEE Trans. Wirel. Commun.*, vol. 10, no. 1, pp. 33–36, Jan. 2011.

9. Y. Zou, Y. Yao, and B. Zheng, "Cognitive transmissions with multiple relays in cognitive radio networks," *IEEE Trans. Wirel. Commun.*, vol. 10, no. 2, pp. 648–659, Feb. 2011.

10. M. Xie, W. Zhang, and K. Wong, "A geometric approach to improve spectrum efficiency for cognitive relay networks," *IEEE Trans. Wirel. Commun.*, vol. 9, no. 1, pp. 268–281, Jan. 2010.

11. N. Devroye, P. Mitran, and V. Tarokh, "Achievable rates in cognitive radio channels," *IEEE Trans. Inf. Theory*, vol. 52, no. 5, pp. 1813–1827, May 2006.

12. J. Zhang, and Q. Zhang, "Stackelberg game for utility-based cooperative cognitive radio networks," *in Proc. MobiHoc*, New Orleans, LA, May 2009.

13. O. Simeone, I. Stanojev, S. Savazzi, Y. Bar-Ness, U. Spagnolini, and R. Pickholtz. "Spectrum leasing to cooperating secondary ad hoc networks," *IEEE J. Sel. Areas Commun.*, vol. 26, no. 1, pp. 203–213, Jan. 2008.

14. M. Levorato, O. Simeone, U. Mitra, and M. Zorzi, "Cooperation and coordination in cognitive networks with packet retransmission," in *Proc. IEEE Int'l Inf. Theory Workshop*, Taormina, Oct. 2009.

15. O. Simeone, Y. Bar-Ness, and U. Spagnolini, "Stable throughput of cognitive radios with and without relaying capability," *IEEE Trans. Wirel. Commun.*, vol. 55, no. 12, pp. 2351–2360, Dec. 2007.

16. Y. Yi, J. Zhang, Q. Zhang, T. Jiang, and J. Zhang, "Cooperative communication-aware spectrum leasing in cognitive radio networks," in *Proc. DySPAN*, Singapore, Apr. 2010.

17. W. Su, J. Matyjas, and S. Batalama, "Active cooperation between primary users and cognitive radio users in cognitive ad-hoc networks," in *Proc. IEEE ICASSP*, Dallas, TX, Mar. 2010.

18. A. Attar, M. Nakhai, and A. Aghvami, "Cognitive radio game for secondary spectrum access problem," *IEEE Trans. Wirel. Commu.*, vol. 8, no. 4, pp. 2121–2131, Apr. 2009.

19. P. Lin, J. Jia, Q. Zhang, and M. Hamdi, "Dynamic spectrum sharing with multiple primary and secondary users," in *Proc. IEEE Int'l Conf. Commun.* , Cape Town, South Afica, May 2010.

20. Y. Chen, and K. Liu, "Indirect reciprocity game modelling for cooperation stimulation in cognitive networks," *IEEE Trans. Wirel. Commu.*, vol. 59, no. 1, pp. 159–168, Jan. 2011.

21. A. El-Sherif, A. Sadek, and K. Liu, "Opportunistic multiple access for cognitive radio networks," *IEEE J. Sel. Areas Commun.*, vol. 29, no. 4, pp. 704–715, Apr. 2011.

22. S. Hua, H. Liu, M. Wu, and S. Panwar, "Exploiting MIMO antennas in cooperative cognitive radio networks," in *Proc. INFOCOM*, pp. 2714–2722, Shanghai, China, Apr. 2011.

Chapter 3
Orthogonally Dull-Polarized Antenna Based Cooperative Cognitive Radio Networking

Abstract This chapter is concerned with enhancement of spectrum efficiency/utilization by using polarization enabled two-phase cooperation between primary users (PUs) and secondary users (SUs) for cooperative cognitive radio networking (CCRN). Specifically, we aim to exploit the degrees of freedom provided by orthogonally dual-polarized antennas (ODPAs) to attain an interference-free two-phase cooperation framework. The use of ODPAs enables concurrent transmissions of multiple independent signals of PUs and SUs, and interference suppression via polarization zero-forcing and polarization filtering to obtain significant performance improvement. By leveraging both temporal and polarization domains, a polarization based two-timescale CCRN scheme to improve spectrum efficiency/utilization is presented. To maximize a weighted sum throughput of PUs and SUs under energy/power constraints, the problem is formulated and solved based on a multi-timescale Markov decision process, and two modified backward iteration algorithms are devised to attain the optimal policies. Numerical and simulation results validate the effectiveness of the proposed framework for CCRN, showing that the obtained policy outperforms both greedy and random ones.

During the past two decades, we have witnessed an explosive proliferation of applications in wireless communications and networking. State-of-the-art technologies associated with products and services are changing our life styles from various aspects. To support sustainable development of wireless communications and networks, the demand for radio spectrum has been skyrocketing. Since the amount of usable spectrum is finite, frequency bands and their usage are strictly managed and enforced by government regulators. Under this regulatory enforcement, spectrum is statically and exclusively allocated to dedicated networks on a license basis, i.e., only licensed users, also referred to as primary users (PUs), can access the assigned spectrum. However, the legacy fixed spectrum access leads to significant spectral underutilization owing to the sporadic use of spectrum [1, 2]. To address spectral underutilization, dynamic spectrum access has been proposed [2–6]. The methodology of dynamic spectrum access is to enable unlicensed users, referred to as secondary users (SUs), to opportunistically use licensed spectrum without causing harmful interference to the PUs. In this context, SUs can (i) dynamically

© The Author(s) 2016

B. Cao et al., *Cooperative Cognitive Radio Networking*, SpringerBriefs in Electrical and Computer Engineering, DOI 10.1007/978-3-319-32881-2_3

sense and use unused licensed bands, (ii) concurrently access licensed spectrum with PUs on a non-interfering basis, or (iii) cooperatively negotiate with PUs for transmission opportunities through providing tangible service to PUs, e.g., serving as relays for PUs. One key enabling technique for dynamic spectrum access is the cognitive radio (CR) equipped by SUs. A network consists of CRs is referred to as a cognitive radio network. These models are termed interweave cognitive radio networking (CRN), underlay CRN, and overlay CRN, respectively [7].

It is recognized that spectrum sensing is one feasible approach to implementing dynamic spectrum access [4–6]. However, there is no connection between PUs and SUs in both interweave and underlay modes, i.e., SUs are transparent to PUs. Mutual benefit based cooperation between PUs and SUs is a promising way to implement dynamic spectrum access. In this framework, PUs and SUs dynamically select appropriate partners for cooperative relaying to create a win-win situation. The rationale behind this framework is that PUs have the need to improve primary transmissions and SUs have the need to access spectrum for secondary transmissions. Furthermore, the cooperation mechanism can be an alternative way for SUs to access spectrum if no idle spectrum holes can be detected in the spectrum sensing based framework.

The focus of this chapter is on exploiting user cooperation between PUs and SUs to attain overlay CRN, also known as cooperative cognitive radio networking (CCRN). In CCRN, a PU is free to select one or more SUs as relays, and SUs obtain transmission opportunities from the PU as rewards. Considerable cooperation frameworks have been proposed for CCRN, such as (i) three-phase time division multiple access based CCRN [8–10]: a PU broadcasts in the first phase, SUs forward the PU's signal in the second phase, and relaying SUs transmit their own data in the third phase [11, 12]; (ii) two-phase frequency division multiple access based CCRN [11]: a PU uses a large fraction of its licensed band for two-phase communications with SUs, and yields the remaining band to relaying SUs as rewards; (iii) two-phase space division multiple access based CCRN [13, 14]: an SU exploits beamforming provided by spatially distributed multiple antennas to avoid interference on PUs and other SUs, so that the SU can relay the information of the PU while concurrently accessing the same spectrum; and (iv) two-phase quadrature signaling based CCRN [12, 13, 15, 16]: by exploiting two-dimensional modulation in CCRN, an SU is able to relay the PU's data while transmitting its own signal orthogonally in the same time slot.

All the above-mentioned schemes deal with cooperation within a single frame in which the channel is assumed to experience spatial flat fading. Both SUs and PUs aim to optimize the cooperation performance in a greedy manner. However, for an energy-constrained network such as an energy-harvesting sensor network where the battery energy of both SUs and PUs is limited, these schemes can no longer be optimal. Another challenging issue is that each cooperation should be finalized within one frame, while a new relay selection procedure is required at the beginning of the next frame. A trade-off between the cooperative diversity and relay selection overhead should be addressed.

To tackle the aforementioned challenging issues, we propose to utilize the polarization property of electromagnetic waves in CCRN. In the system networked by PUs and SUs, mobile primary transmitters (PTs) communicate with their intended primary receivers (PRs) (e.g., primary base stations), and mobile secondary transmitters (STs) need to communicate with secondary receivers (SRs) (e.g., secondary access points). Both PTs and STs are energy-constrained. SUs are able to avoid interference to PUs and other cooperating SUs by only leveraging polarization filtering and polarization zero-forcing capabilities provided by orthogonally dual-polarized antennas (ODPAs), given that PUs still equip with legacy uni-polarized antennas. The other reason that we introduce polarization in CCRN is due to the limitation of MIMO based frameworks in some applications with physical size constraint. Although the MIMO technology can offer significant enhancement in throughput and link range without additional bandwidth and/or transmit power, the application of MIMO is limited by the hardware cost and physical size. In theory, typical antenna elements must be spaced at least half a wavelength at the mobile terminal and ten wavelengths at the base station or the access point to ensure fully independent spatial channel fading. Therefore, the capability and degrees of freedom provided by MIMO systems are determined by sufficient space interval of antenna elements. This results in a large hardware size of SUs.

One obvious advantage of using ODPA is that there is no spacing requirement for the two antennas since they are co-located, which makes ODPA more suitable for devices with size and cost limitation [18]. This enables SUs to cooperatively relay the data for PUs while concurrently accessing the same spectrum to transmit their own information in cost- and space- efficient manners. Another byproduct of using ODPA is the fact that polarization fading in terms of cross-polar discrimination (XPD) changes more slowly than its spatial counterpart, multi-path fading [19]. The use of ODPA enables SUs to keep using the same polarization states to attain interference suppression in a large time-scale. This indicates that polarization based scheme is more stable than MIMO based scheme with respect to (w.r.t.) time variations. It is noted that there are two time-scales in the proposed framework: The small frame corresponds to the spatial multi-path fading as in [8–11, 14–16], while the large frame called superframe corresponds to the polarization fading. Each superframe consists of N frames, where N is an integer. After selecting appropriate SUs as cooperators, a PU can perform a consecutive N-frame cooperation with the same SUs. In this context, an energy-constrained multi-timescale cooperation framework is proposed. Specifically, at the beginning of each superframe, the PU selects the most effective SUs as cooperators, and determines its transmission power for the current superframe and the cooperation duration of one frame according to long-term channel statistics and residual energy in the previous superframe. Within a superframe, SUs determine the optimal transmission and relay power for each frame according to instantaneous channel coefficients and residual energy in the previous frame. The corresponding resource allocation problem is formulated as a multi-timescale dynamic programming in terms of maximizing a weighted sum throughput subject to power constraints. The optimal policies of the frame- and superframe-level resource allocation problems are proved and analyzed. Modified backward iteration algorithms and the associated numerical and simulations validate the effectiveness of the proposed scheme, showing that the proposed policy outperforms both greedy and random power policies. The main contents of this chapter are four-fold.

1. Presentation of a two-phase cooperation framework for polarization based CCRN: By exploiting the degrees of freedom and leveraging the polarization filtering and polarization zero-forcing provided by ODPA, SUs are able to cooperatively relay the PU's packet and transmit their own signal without causing interference. To the best of our knowledge, this is the first work that introduces polarization to CCRN, and we firstly propose to use polarization zero-forcing to suppress interference;
2. Construction of a multi-timescale frame structure for long-term cooperation: Due to the long-term characteristic of polarization fading w.r.t. time, a superframe consisting of multiple frames is constructed. In this way, PUs and SUs can keep long-term cooperation, so that the trade-off between cooperation gain and selection overhead can be attained;
3. Quantification of the mutual benefit for the PU and SUs in terms of achievable maximum weighted sum throughput: To achieve the mutual benefit, we formulate a weighted sum throughput optimization problem under power constraints in polarization based CCRN;
4. Formulation and solution of the resource allocation problem: Based on the proposed two-phase cooperation framework, the resource allocation problem is formulated as a multi-timescale dynamic programming problem. By decomposing the original problem into two subproblems, the optimal policy for the frame-level resource allocation problem is proved. Then, the superframe-level resource allocation problem is solved by existing policy iteration algorithms. The proposed resource allocation policies are characterized by low computational complexity, while a performance bound (w.r.t. the optimal solution) is derived.

In practical applications, the proposed cooperation framework ensures a high spectrum utilization for size- and energy-constrained SUs in the context of CCRN, which in turn, stimulates the leasing of the licensed spectrum by PUs (Fig. 1.4).

The remainder of the chapter is organized as follows. Section 3.1 describes the system model and fundamentals on polarization with its application in CCRN. In Sect. 3.2, the resource allocation problem is formulated as a two-timescale weighted sum throughput maximization problem. The proposed resource allocation policies and detail analysis are presented in Sect. 3.3. Modified backward iteration algorithms for calculating the frame-level policy, the associated numerical results and performance comparison with greedy and random policies are given in Sect. 3.4, followed by concluding remarks and future work in Sect. 3.5.

Table 3.1 Summary of important symbols

Symbol	Definition
N	Number of frames in one superframe
ϵ	Polarized angel of a dual-polarized signal
δ	Phase difference between two polarized components
$\mathbf{U}_i$	Transmit polarization state of node i
$\mathbf{D}_{ij}$	Depolarization matrix of wireless channel between node i and node j
$\mathbf{h}_{ij}$	Instantaneous channel coefficient between node i and node j
$\mathbf{x}_i(t)$	Transmitted signal of node i
$\mathbf{y}_i(t)$	Received signal of node i
$\mathbf{w}$	Two dimensional addictive Gaussian white noise
σ^2	Noise power
$\mathbf{E}_{uv}$	Oblique projection operator from subspace $\mathbf{u}$ onto $\mathbf{v}$
$\mathbf{R}_{\mathbf{u}}^{\perp}$	Orthogonal projection operation onto subspace $\mathbf{u}$
$\mathbf{I}$	Unit matrix
$\mathbf{d}_i$	Composite vector of depolarization and spatial channel fading of node i
Ω	Time duration of multi-user coordination phase in one superframe
T	Time duration of cooperative transmissions in one frame
ω	Time duration of channel sensing in one frame
β	Time allocation coefficient of cooperative communications in one frame
$P_{P,m}$	PU's transmit power in the mth frame
$P_{STA,m}^{(n)}$	ST_A's transmit power in the mth frame of the nth superframe
$P_{SRA,m}^{(n)}$	SR_A's transmit power in the mth frame of the nth superframe
$P_{STBP,m}^{(n)}$	ST_B's relaying power in the mth frame of the nth superframe
$P_{STBS,m}^{(n)}$	ST_B's power for its own transmission in the mth frame of the nth superframe
$R_{SP,m}^{(n)}$	Achievable channel capacity of the PU obtained from ST_B's relaying path
$T_{P,m}^{(n)}$	Achievable throughput of the PU in the mth frame of the nth superframe
$X_{i,m}^{(n)}$	Initial energy of node i in the mth frame of the nth superframe
ζ_m	Weighting parameter of weighted sum throughput
T_{WSum}	Weighted sum throughput of PU and SUs
γ_m	Polarization and fading profile of the mth frame
$\bar{\mathscr{P}}_{STBS}$	Class of all admissible policies
η	Discount factor
V_{WSum}^*	Value function of weighted sum throughput
Q_n	Objective function

3.1 System Model

3.1.1 Networking Architecture and System Description

Consider a system consists of both PUs and SUs. Mobile PTs communicate with PRs, and mobile STs need to communicate with SRs, while SUs are net-worked in a CCRN fashion. PTs and STs are battery-powered so that they are energy-constrained, while PRs and SRs are AC-powered, i.e., there is no energy constraint for them. In addition, PUs and SUs transceivers work at half-duplex mode, i.e., they cannot transmit and receive simultaneously, and SUs use the decode-and-forward mode to relay PUs' packets. The cooperators fully trust each other, i.e., there are no malicious users and misbehaviors after the cooperation is established. Furthermore, it is assumed that SUs utilize the capability provided by ODPAs in terms of vertical (V) polarization and horizontal (H) polarization to access PUs' spectrum for secondary transmissions coexisting with associated primary transmissions, or to cooperatively relay PUs' data while concurrently transmitting SUs' information in licensed bands. The PUs still use legacy devices and is not required to change their hardware to support dual-polarized capability, e.g., each PU is equipped with the traditional vertically uni-polarized antenna. In this context, the links between SUs, from SUs to PUs, from PUs to SUs, and between PUs are VH to VH (VH2VH), VH2V, V2VH, and V2V transmissions in terms of polarization, respectively, where VH2VH means the transmitter and receiver both have ODPA antennas in terms of vertical and horizontal polarization; VH2V means the transmitter is with vertical/horizontal ODPA and the receiver is with a single vertically polarization antenna, and vice versa for V2VH; and V2V means the transmitter and receiver are both with single vertically polarized antennas, respectively.

3.1.2   Representation of Polarization

In a right-handed x–y–z *Cartesian* coordinate system with the z-coordinate representing the propagating direction, and the x-coordinate and y-coordinate representing H- and V-polarized vectors, the two-dimensional signal $\mathbf{x}(t)$ radiated by ODPA can be expressed in the form of Jones vector

$$\mathbf{x}(t) = \begin{bmatrix} x_H(t) \\ x_V(t) \end{bmatrix} = \begin{bmatrix} \cos \varepsilon \\ \sin \varepsilon \exp (j\delta) \end{bmatrix} x(t) = \mathbf{U}x(t) \qquad (3.1)$$

where $x(t)$ is the waveform of $\mathbf{x}(t)$, ε is the polarized angle which denotes amplitude relationship between V- and H-polarized components, i.e., $\varepsilon = \arctan(|x_V(t)|/|x_H(t)|)$, and δ describes phase difference between them, i.e., $\delta = \arg\{x_V(t)\} - \arg\{x_H(t)\}$, and $\mathbf{U} = [\cos \varepsilon, \sin \varepsilon \exp(j\delta)]^T$ is the transmitted polarization state of $\mathbf{x}(t)$.

Due to imperfect antenna cross-polar isolation in ODPA and cross-polar ratio caused by propagation mediums [18], the effect of depolarization which is determined by these two factors, should be considered if polarization is to be exploited. The degree of depolarization is described by a cross-polar discrimination (XPD), and XPD varies slowly comparing with spatial multi-path fast fading when signals propagate in the wireless channel [18, 19]. For simplicity, it is further assumed that

ODPA is comprised of linearly polarized antennas, so that cross-polar isolation and cross-polar ratio can be decoupled. The ODPA used in this chapter is with infinite cross-polar isolation, i.e., an ideal polarization isolation between two antennas, then cross-polar ratio is the only factor determines XPD. Therefore, the depolarization effect can be described by a 2×2 matrix $\mathbf{D}$ [18], given by

$$\mathbf{D} = \begin{bmatrix} D_{HH}, & D_{HV} \\ D_{VH}, & D_{VV} \end{bmatrix}. \tag{3.2}$$

Furthermore, we have $D_{HV} = D_{HH} = 0$ for VH2V transmissions and $D_{VH} = D_{HH} = 0$ for V2V and V2VH transmissions. Throughout this chapter, the XPD is assumed to be known by SUs. Interested readers can refer to [19] for more details about XPD estimation.

3.1.3 Fundamentals of Exploiting Polarization for CCRN

Since secondary transmissions coexist with primary transmissions in the same frequency band and time slot, co-channel interference from SUs to PUs should be avoided. Moreover, SUs need to acquire secondary and primary signals separately from the mixed signals, as SUs need to obtain their own signals while forwarding PUs' signals.

Due to the independence with frequency, space, time, and code domains, the polarization processing can be an alternative to achieve interference avoidance. As one feasible application of polarization processing to suppress co-channel interference, polarization filtering attracts a great interest in recent decades, e.g., polarization filtering has been widely studied and applied in radar and communication systems [20–24]. While polarization filtering attracts little attention in wireless communications, we propose to introduce polarization filtering into CCRN to expand available degrees of freedom. The main principle of polarization filtering is based on the assumption that the polarization state of the interference is different from that of the desired signal. The authors proposed the oblique projection polarization filtering technique to suppress the interference that is not orthogonal with the desired signal in the polarization domain [21, 22]. In this chapter, oblique projection polarization filtering is used by SUs to separate secondary and primary signals [21, 22], and polarization zero-forcing is used to suppress secondary signals at the PR [23]. The principle of polarization zero-forcing is that SUs can choose their polarization states according to PUs depolarization matrices, so that SUs' signals are zero-forced automatically at PR in the polarization domain.

To present the feasibility of using oblique projection polarization filtering and polarization zero-forcing to avoid co-channel interference, how to exploit oblique projection polarization filtering and polarization zero-forcing in CCRN for one PU and one SU case is investigated [17]. As shown in Fig. 3.1, PT sends signal $x_P(t)$ to PR, and ST sends $x_S(t)$ to SR in the same frequency band with PT.

Fig. 3.1 Exploiting PF into
CRN: one PU and one SU
case

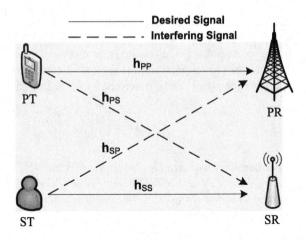

The polarization states used by PT and ST are denoted by $\mathbf{U}_P = [0, \pm 1]^T$ and $\mathbf{U}_S = [\cos \varepsilon_P, \pm \sin \varepsilon_P]^T$, respectively. According to (3.1), the received signal $\mathbf{y}_S(t)$ at SR can be expressed as

$$\mathbf{y}_S(t) = \mathbf{h}_{SS}\mathbf{U}_S x_S(t) + \mathbf{h}_{PS}\mathbf{U}_P x_P(t) + \mathbf{w} \tag{3.3}$$

where $\mathbf{h}_{SS} = h_{SS}\mathbf{D}_{SS}$ (resp. $\mathbf{h}_{PS} = h_{PS}\mathbf{D}_{PS}$) is the composite channel fading matrix containing spatial fading coefficient h_{SS} (resp. h_{PS}) and polarization fading in terms of depolarization matrix $\mathbf{D}_{SS}$ (resp. $\mathbf{D}_{PS}$) of ST to SR (resp. PT to SR). Matrix $\mathbf{w}$ represents the two-dimensional additive noise with zero mean, and covariance matrix $\sigma^2\mathbf{I}$.

Define $\mathbf{D}_{SS}\mathbf{U}_S = \mathbf{d}_S$ and $\mathbf{D}_{PS}\mathbf{U}_P = \mathbf{d}_P$, the oblique projection operator that projects vectors onto subspace spanned by $\langle \mathbf{d}_S \rangle$ along the subspace spanned by $\langle \mathbf{d}_P \rangle$ can be written as [25]

$$\mathbf{E}_{\mathbf{d}_S \mathbf{d}_P} = \mathbf{d}_S(\mathbf{d}_S^H \mathbf{R}_{\mathbf{d}_P}^\perp \mathbf{d}_S)^{-1}\mathbf{d}_S^H \mathbf{R}_{\mathbf{d}_P}^\perp \tag{3.4}$$

where $\mathbf{R}_{\mathbf{d}_P}^\perp = \mathbf{I} - \mathbf{d}_P(\mathbf{d}_P^H \mathbf{d}_P)^{-1}\mathbf{d}_P^H$ is the orthogonal projection operator which projects vectors onto the complementary subspace spanned by $\langle \mathbf{d}_P \rangle$. It is noted that there is no requirement on orthogonality between $\mathbf{d}_P$ and $\mathbf{d}_S$. Since $\mathbf{R}_{\mathbf{d}_P}^\perp \mathbf{d}_P = 0$, and this further leads to $\mathbf{E}_{\mathbf{d}_S \mathbf{d}_P}\mathbf{d}_S = \mathbf{d}_S$ and $\mathbf{E}_{\mathbf{d}_S \mathbf{d}_P}\mathbf{d}_P = 0$. In this way, the following result holds true

$$\mathbf{E}_{\mathbf{d}_S \mathbf{d}_P}\mathbf{y}_S(t) = \mathbf{h}_{SS}\mathbf{U}_S x_S(t) + \mathbf{E}_{\mathbf{d}_S \mathbf{d}_P}\mathbf{w}. \tag{3.5}$$

Along the same analysis, the SU can extract the PU's signal by using $\mathbf{E}_{\mathbf{d}_P \mathbf{d}_S}$ which is the operator that obliquely projects vectors onto subspace spanned by $\langle \mathbf{d}_P \rangle$ along the subspace spanned by $\langle \mathbf{d}_S \rangle$. The PU's signal can be obtained by $\mathbf{E}_{\mathbf{d}_P \mathbf{d}_S}\mathbf{y}_S(t)$.

Therefore, the SU can separate its own desired signal and the PU's signal by using oblique projection polarization filtering.

To avoid the interference from SUs to PU, polarization zero-forcing is used. The signal received by PR is given by

$$y_P = \{h_{PP}\mathbf{D}_{PP}\mathbf{U}_P x_P + h_{SP}\mathbf{D}_{SP}\mathbf{U}_S x_S\}|_V + w \quad (3.6)$$

where $\{\mathbf{x}(t)\}|_V$ is the V-polarized component of $\mathbf{x}(t)$, and w is the 1-D additive noise with zero mean and variance σ^2. If $\mathbf{D}_{SP}\mathbf{U}_S = 0$ is achieved, SU's signal is zero-forced at PR in the polarization domain. Therefore, after getting $\mathbf{D}_{SP}$, the SU can set its polarization state accordingly. It is noted that the proposed polarization zero-forcing is similar to its counterpart beamforming in MIMO. Beamforming uses spatial CSI, while polarization zero-forcing uses XPD, and XPD changes more slowly than spatial CSI w.r.t. time. In this way, polarization zero-forcing is more stable than beamforming w.r.t. time, and there is no spacing requirement for ODPA placement.

Based on the polarization processing, the SU's signal does not interfere with primary transmission without changing PU's transceiver design by using polarization zero-forcing, and the SU's signal can be separated from the PU's signal by using oblique projection polarization filtering.

3.1.4 Multi-Timescale Cooperation Framework

As shown in Fig. 3.2a, time is divided into superframes. A PU keeps cooperating with the same SUs for one superframe with length $\Omega + N(T + \omega)$, i.e., the PU changes its cooperators on a superframe basis. Each superframe is further divided into two parts, the first part is the multi-user coordination process with duration Ω, i.e., the PU selects the most effective SUs for cooperation. The second part is the cooperation process with length $N(T + \omega)$, and this process is composed of N consecutive frames, where N is a positive integer. Constant power is used by a PU to cooperate with SUs within each superframe.

Since polarization fading in terms of XPD changes much slower as compared with spatial multi-path fading, the duration of each superframe is chosen such that XPDs remain constants. The length of $T + \omega$ is determined by Doppler spread w.r.t. licensed spectrum and mobility, e.g., in duration $T + \omega$, channel suffers flat-fading, and channel coefficients are constants. Because channel coefficients and residual energy are different from frame to frame, each SU should adapt transmitting and relaying power in different frames to achieve satisfactory performance.

For the network scenario under consideration, we investigate two ST-SR pairs simultaneously cooperate with one PT-PR pair during one cooperation period. Each frame in one superframe is divided into two parts, as shown in Fig. 3.2b. The first part with duration ω is used to estimate channel coefficients of each frame, while the second part is used for two-phase cooperation. Consider one specific frame

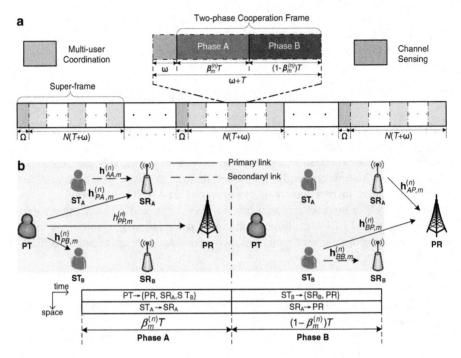

Fig. 3.2 System model, superframe and frame structures of the proposed CCRN. (**a**) The PU's cooperation process and the structure of a superframe. (**b**) Structure of two-phase cooperative communications in each frame

within the mth superframe. In the first cooperation phase with duration $\beta_m T$, say phase A, PT and ST_A transmit signal to PR and SR_A, respectively. At the same time, SR_A and ST_B receive signals from both ST_A and PT. For PR and ST_B, the signal from ST_A is considered as interference. However, by using oblique projection polarization filtering, ST_A's signal can be nulled at PR, and ST_B can extract primary signal from mixed signals, while SR_A can separate its own desired signal and primary signal. In the second phase with duration $(1 - \beta_m)T$, say phase B, ST_B cooperatively relays PT's information received in phase A while concurrently sends its own signals to SR_B. Meanwhile, SR_A forwards the primary information to PR. Based on the cooperation framework, the throughput achieved within each phase can be calculated as follows:

Phase A: In phase A of the nth frame, PT sends 1-D signal $x_P^{(n)}$ to PR with constant power $P_{P,m}$. Meanwhile, ST_A sends 2-D signal $\mathbf{x}_{SA}^{(n)} = \mathbf{U}_{STA,m}x_{SA}^{(n)}$ to SR_A with power $P_{STA,m}^{(n)}$. Due to the broadcast nature and half-duplex mode, PR, SR_A and ST_B can hear PT and ST_A's signals. By introducing the proposed oblique projection polarization filtering and polarization zero-forcing at SUs, co-channel interference can be suppressed, e.g., PT can only get the primary signal, SR_A can separate secondary and primary signals and ST_B can extract the primary signal.

The achievable rate of primary and secondary links in phase A of the nth frame can be obtained as

$$R_{PS,m}^{(n)} = S\left(\frac{\min_{i\in\{A,B\}}\{\sin^2 \gamma_{Pi,m}|h_{Pi,m}^{(n)}|^2\}P_{P,m}}{\sigma^2}\right)$$

and

$$R_{SA,m}^{(n)} = S\left(\frac{\sin^2 \gamma_{PA,m}|h_{AA,m}^{(n)}|^2 P_{STA,m}^{(n)}}{\sigma^2}\right),$$

where $S(x) = \log_2(1 + x)$ is the Shannon capacity when signal-to-noise ratio is x, $h_{pq,m}^{(n)}$ is the channel coefficient between nodes p and q in the mth superframe, and $\gamma_{Pi,m}$ ($i \in \{A, B\}$) is determined by the angle of polarization vectors between PT and SUs, e.g., $\gamma_{PA,m} = \arccos(\mathbf{d}^H\mathbf{s})$, this results from the fact that the additive noise is amplified after oblique projection onto one subspace along an oblique subspace [25], and the amplification factor is determined by the principal angle between two subspaces.[1]

Phase B: SR_A forwards primary information to PR with power $P_{SRA,m}^{(n)}$, and ST_B forwards primary information to PR using power $P_{STBP,m}^{(n)}$ and concurrently sends its own information to SR_B with power $P_{STBS,m}^{(n)}$. By using oblique projection polarization filtering and polarization zero-forcing, there is no interference between PUs and SUs. PR uses maximal ratio combining to combine signals received from different paths. Therefore, PU and SU_B's achievable rate are given by

$$R_{SP,m}^{(n)} = S\left(\frac{|h_{AP,m}^{(n)}|^2 P_{SRA,m}^{(n)}}{\sigma^2} + \frac{|h_{BP,m}^{(n)}|^2 P_{STBP,m}^{(n)}}{\sigma^2}\right)$$

and

$$R_{SB,m}^{(n)} = S\left(\frac{|h_{BB,m}^{(n)}|^2 P_{STBS,m}^{(n)}}{\sigma^2}\right),$$

respectively.

Considering the transmissions in both phase A and phase B, the achievable primary and secondary throughput in the nth frame can be obtained by

---

[1]For additive noise at SR_A, the noise variance after oblique projection is $\frac{\sigma^2}{\sin^2 \gamma_{PA,m}}$. Especially, if polarization vectors $\mathbf{d}$ and $\mathbf{s}$ are orthogonal, then $\gamma_{PA,m} = 90°$ which means noise is invariant after projection.

$T_{P,m}^{(n)} = \min\{\beta_m TR_{PS,m}^{(n)}, (1 - \beta_m)TR_{SP,m}^{(n)}\}$, $T_{SA,m}^{(n)} = \beta_m TR_{SA,m}^{(n)}$, and $T_{SB,m}^{(n)} = (1 - \beta_m)$ $TR_{SB,m}^{(n)}$, respectively. Assume that the PU knows the long-term channel fading information such as the first and the second-order statistics of itself and SUs, and denote $T_{P,m}$, $R_{PS,m}$, and $R_{SP,m}$ as the throughput of the PU in the mth superframe, the throughput of PT to SUs and that of SUs to PR during the mth superframe. The optimal β_m for the mth superframe can be obtained by solving $\beta_m R_{PS,m} = (1 - \beta_m)R_{SP,m}$.

The energy consumptions of PT, SU_A, and SU_B in the nth frame of the mth superframe can be calculated as $\beta_m TP_{P,m}$, $\beta_m TP_{STA,m}^{(n)}$, and $(1 - \beta_m)T(P_{STBP,m}^{(n)} + P_{STBS,m}^{(n)})$, respectively. Suppose the residual energy of PT, ST_A, and ST_B at the beginning of the nth frame in the mth superframe is $X_{P,m}^n$ ($X_{P,m}^n \in [0, X_{PM}]$), $X_{STA,m}^n$ ($X_{STA,m}^n \in [0, X_{STAM}]$), and $X_{STB,m}^n$ ($X_{STB,m}^n \in [0, X_{STBM}]$), respectively, where X_{PM}, X_{STAM}, and X_{STBM} are the battery capacities. Then, the residual energy of PT, ST_A, and ST_B in the $(n + 1)$th frame (i.e., after the transmissions in the nth frame) is, respectively, given by

$$X_{P,m}^{(n+1)} = \left[X_{P,m}^{(n)} - \beta_m TP_{P,m}\right]^+,$$

$$X_{STA,m}^{(n+1)} = \left[X_{STA,m}^{(n)} - \beta_m TP_{STA,m}^{(n)}\right]^+, \tag{3.7}$$

$$X_{STB,m}^{(n+1)} = \left[X_{STB,m}^{(n)} - (1 - \beta_m)T(P_{STBP,m}^{(n)} + P_{STBS,m}^{(n)})\right]^+,$$

where $[X]^+$ equals X if $X > 0$ and 0 otherwise. Since the maximum volume of data that can be delivered to PR is $\beta_m TR_{PS,m}^{(n)}$, the relaying power $P_{STBP,m}^{(n)}$ used by ST_B is further bounded by $P_{STBP,m}^{(n)} \leq \left[\left(1 + \frac{\min_{i \in \{A,B\}} \sin^2 \gamma_{Pi,m}|h_{Pi,m}^{(n)}|P_{P,m}}{N_0}\right) \frac{N_0 2^{\frac{\beta_m}{1-\beta_m}}}{|h_{BP,m}^{(n)}|^2} - 1\right]^+ \overset{\triangle}{=}$ $P_{STBPM',m}^{(n)}$. Taking into account the transmission power limitation of ST_B, we have $P_{STBP,m}^{(n)} \leq \min\{P_{STBM}, P_{STBPM',m}^{(n)}\} \overset{\triangle}{=} P_{STBPM,m}^{(n)}$.

The weighed sum throughput in the nth frame of the mth superframe is given by $T_{WSum,m}^{(n)} = (1 - \zeta_m)T_{P,m}^{(n)} + \zeta_m(T_{SA,m}^{(n)} + T_{SB,m}^{(n)})$, where ζ_m is the weight of $T_{WSum,m}^{(n)}$. Obviously, $T_{WSum,m}^{(n)}$ is a generalized form. For $\zeta_m = 0$, $T_{WSum,m}^{(n)}$ represents the primary throughput, and for $\zeta_m = 1$, $T_{WSum,m}^{(n)}$ is secondary throughput, while for $\zeta_m = \frac{1}{2}$, $T_{WSum,m}^{(n)}$ is equivalent to the total throughput. As a result, a balance between primary and secondary throughput can be achieved by adjusting ζ_m. If the primary link is totally blocked, then SUs offer multi-hop service to the PU. In this event, the PU is willing to agree on a large ζ_m for cooperation with SUs. In this chapter, ζ_m is pre-defined and is based on the agreement between the PU and SUs.

3.2 Problem Formulation

3.2.1 Definition of the Resource Allocation Policy

Based on the system model and frame structure discussed above, the resource allocation in terms of power allocation should be performed in two timescales. Specifically, at the beginning of each superframe, the selection of cooperating SUs and the calculation of $P_{P,m}$ are performed based on the residual energy of the PU and the polarization and long-term (for one superframe) fading statistics of SUs. Then, the selected SUs perform resource allocation according to the resource allocation policy at the frame level. Within each frame, ST_A selects the power for transmission to SR_A (i.e., $P_{STA,m}^{(n)}$), while ST_B selects the transmission power for both P_R and SR_B (i.e., $P_{STBP,m}^{(n)}$ and $P_{STBS,m}^{(n)}$, respectively).

In details, at the beginning of the mth superframe, denote the energy level of the PU as $X_{P,m}^{(0)} \in [0, X_{PM}]$. The polarization and long-term channel fading characteristics (w.r.t. all SUs) are denoted as Γ_m. Based on the energy level and fading statistics, the PU selects ST_A and ST_B and determines its own transmission power $P_{P,m}$. Suppose the resource allocation policy is a function which maps the energy level of the PU and the polarization and long-term channel fading characteristics to the two selected SUs, the transmission power is given by

$$P_{P,m} : [0, X_{PM}] \times \Re \to \mathscr{ST} \times \mathscr{ST} \times [0, P_{PM}] \tag{3.8}$$

where $\Re$ is the set of all possible values of Γ_m, $\mathscr{ST}$ is the set of all SUs, and P_{PM} is the maximum transmission power of the PU.

At the beginning of the nth frame in the mth superframe, the polarization and fading profiles are denoted as $\gamma_m = (\gamma_{Pi,m}(i \in \{A, B\}), \gamma_{STA,m}, \gamma_{STB,m})$ and $h_m^{(n)} = (h_{Pi,m}^{(n)}, h_{STA,m}^{(n)}, h_{STB,m}^{(n)})$, respectively. Given the energy level ($X_{STA,m}^{(n)} \in [0, X_{STAM}]$ and $X_{STB,m}^{(n)} \in [0, X_{STBM}]$) and CSI (including polarization and fading profiles) of the PU and SUs, the resource allocation policies of ST_A and ST_B are given by

$$P_{STA,m}^{(n)} : [0, X_{STAM}] \times \mathbb{R}_+^3 \times \mathbb{R}_+^3 \to [0, P_{STAM}] \tag{3.9}$$

$$P_{STB,m}^{(n)} : [0, X_{STBM}] \times \mathbb{R}_+^3 \times \mathbb{R}_+^3 \to [0, P_{STBPM,m}^{(n)}] \times [0, P_{STBSM}] \tag{3.10}$$

where the first and second decision variables in (3.10) correspond to the transmission power for primary and secondary information of ST_B, respectively. Let $\bar{\mathbf{P}}_P = \{P_{P,m} | m \in \{0, 1, 2, \cdots, M - 1\}\}$, $\bar{\mathbf{P}}_{STA} = \{P_{STA,m}^{(n)} | m \in \{0, 1, 2, \cdots, M - 1\}$, $n \in \{0, 1, 2, \cdots, N - 1\}\}$, $\bar{\mathbf{P}}_{STBP} = \{P_{STBP,m}^{(n)} | m \in \{0, 1, 2, \cdots, M - 1\}, n \in \{0, 1, 2, \cdots, N - 1\}\}$, and $\bar{\mathbf{P}}_{STBS} = \{P_{STBS,m}^{(n)} | m \in \{0, 1, 2, \cdots, M - 1\}, n \in \{0, 1, 2, \cdots, N - 1\}\}$, and let $\bar{\mathscr{P}}_P$, $\bar{\mathscr{P}}_{STA}$, $\bar{\mathscr{P}}_{STBP}$, and $\bar{\mathscr{P}}_{STBS}$ be the class of all admissible policies for the PU, ST_A and ST_B, respectively.

3.2.2 Resource Allocation Problem Formulation

The state variables are defined w.r.t. the residual energy of PT, ST_A, and ST_B, i.e., $X_{P,m}^{(n)}$, $X_{STA,m}^{(n)}$, and $X_{STB,m}^{(n)}$, respectively, which evolve according to (3.7). Denote the CSI at the beginning of the nth frame of the mth superframe as $\mathbf{S}_m^{(n)} = (\gamma_m, h_m^{(n)})$. The objective of the resource allocation is to maximize the sum of the weighted sum throughput over the M superframes. Then we have the following optimization problem:

$$\textbf{(P1)} \max_{\tilde{P}_P \in \mathscr{P}_P} \max_{\substack{\tilde{P}_{STA} \in \mathscr{P}_{STA} \\ \tilde{P}_{STBP} \in \mathscr{P}_{STBP}, \tilde{P}_{STBS} \in \mathscr{P}_{STBS}}} \lim_{M \to \infty} \left\{ E \left[\sum_{m=0}^{M-1} \eta^m \sum_{n=0}^{N-1} T_{WSum}(\mathbf{S}_m^{(n)}, P_{P,m}, P_{STA,m}^{(n)}, P_{STBP,m}^{(n)}, P_{STBS,m}^{(n)}) \right] \right\}$$

(3.11)

where $T_{WSum}(\cdot)$ represents the weighted sum throughput of the nth frame in the mth superframe defined in Sect. 3.1. The discount factor η is used to emphasize the short-term reward since the system statistics are more likely to change in a distant future. Note that the number of superframes is considered to be sufficiently large in this scenario, i.e., an approximately infinite horizon problem ($M \to \infty$) without a termination state is investigated.

3.3 Optimal Resource Allocation Policy

According to the theory of Markov decision process (MDP), for the multi-timescale resource allocation problem P1, optimal resource allocation policies in the forms of (3.8), (3.9), and (3.10) exist for PT, ST_A, and ST_B, respectively [26]. However, finding the optimal policies is computationally prohibitive because of the multi-timescale framework. In order to reduce the computational complexity, we consider a transformation of problem P1, based on which the optimal frame-level resource allocation policy can be obtained explicitly, while the superframe-level resource allocation can be achieved by existing policy iteration algorithms with bounded performance guarantee. Denote the value function of problem P1 as

$$V_{WSum}^*(X_P, X_{STA}, X_{STB}, \mathbf{S}) = \max_{\tilde{P}_P \in \mathscr{P}_P} \max_{\substack{\tilde{P}_{STA} \in \mathscr{P}_{STA} \\ \tilde{P}_{STBP} \in \mathscr{P}_{STBP}, \tilde{P}_{STBS} \in \mathscr{P}_{STBS}}} \lim_{M \to \infty} \left\{ \sum_{m=0}^{M-1} \eta^m \sum_{n=0}^{N-1} E \Big[\right.$$

$$\left. T_{WSum}(\mathbf{S}_m^{(n)}, P_{P,m}, P_{STA,m}^{(n)}, P_{STBP,m}^{(n)}, P_{STBS,m}^{(n)}) \Big] \right\}$$ (3.12)

where $X_{P,0}^{(0)} = X_P$, $X_{STA,0}^{(0)} = X_{STA}$, and $X_{STB,0}^{(0)} = X_{STA}$ are the initial battery energy, while $\mathbf{S}$ represents the initial value of CSI. Note that the equality in (3.12) holds since the expectation of summation equals the summation of expectations. By approximating the CSI and the selected SUs as independent among superframes, the value function is given by

$$
\hat{V}^*_{WSum}(X_P, X_{STA}, X_{STB}, \mathbf{S}) = \max_{\bar{\mathbf{P}}_P \in \bar{\mathscr{P}}_P} \left\{ \max_{\substack{\mathbf{P}_{STA} \in \mathscr{P}_{STA} \\ \mathbf{P}_{STBP} \in \mathscr{P}_{STBP}, \mathbf{P}_{STBS} \in \mathscr{P}_{STBS}}} \{R(X_P, X_{STA}, X_{STB}, \mathbf{S})\} \right.
$$
$$
\left. + \eta E_{X'_{STA}, X'_{STB}, \mathbf{S}'} \left[\hat{V}^*_{WSum}(X'_P, X'_{STA}, X'_{STB}, \mathbf{S}') \right] \right\}
$$
$$(3.13)$$

where the expectation is performed w.r.t. the randomness in the available battery energy of newly selected SUs in the next superframe (X'_{STA} and X'_{STB}) and the CSI ($\mathbf{S}'$), while the battery energy of the PU under consideration (from X_P to X'_P) evolves according to (3.7). In (3.13), the resource allocation vectors within a superframe are simplified by removing subscript m (w.r.t. a tagged superframe) and are given by $\mathbf{P}_{STA} = \{P^{(n)}_{STA} | n \in \{0, 1, 2, \cdots, N-1\}\}$, $\mathbf{P}_{STBP} = \{P^{(n)}_{STBP} | n \in \{0, 1, 2, \cdots, N-1\}\}$, and $\mathbf{P}_{STBS} = \{P^{(n)}_{STBS} | n \in \{0, 1, 2, \cdots, N-1\}\}$, respectively. Accordingly, $\mathscr{P}_{STA}$, $\mathscr{P}_{STBP}$, and $\mathscr{P}_{STBS}$ denote the class of all admissible policies for ST_A and ST_B within the tagged superframe. Function $R(\cdot) = \sum_{n=0}^{N-1} E\left[T^{(n)}_{WSum}(\cdot) \right]$ represents the one-superframe reward w.r.t. CSI $\mathbf{S}$ and initial battery energy X_P, X_{STA}, and X_{STB}.

A quantitative performance degradation of the approximation can be evaluated based on Müller's work [26]. However, different from the traditional problems, the frame-level resource allocation is performed w.r.t a finite number of frames (within one superframe). Specifically, based on the specific structure of the cooperation framework, an optimal frame-level resource allocation policy can be explicitly derived, to be discussed as follows. Denote the objective functions of ST_A and ST_B as $T_{WSum,A}(\cdot)$ and $T_{WSum,B}(\cdot)$, respectively. By separating the terms in $T_{WSum}(\cdot)$ w.r.t. ST_A and ST_B, we have

$$
T_{WSum,A}(\mathbf{S}^{(n)}, P^{(n)}_{STA}) = \zeta T_{SA}(\mathbf{S}^{(n)}, P^{(n)}_{STA}) \tag{3.14}
$$

$$
T_{WSum,B}(\mathbf{S}^{(n)}, P_P, P^{(n)}_{STBP}, P^{(n)}_{STBS}) = (1 - \zeta) T_P(\mathbf{S}^{(n)}, P_P, P^{(n)}_{STBP}) + \zeta T_{SB}(\mathbf{S}^{(n)}, P^{(n)}_{STBS}).
$$
$$(3.15)$$

Since the resource allocation of ST_A can be considered as a special case of the resource allocation of ST_B (i.e., one-dimensional power allocation instead of two-dimensional power allocation), we derive the optimal frame-level resource allocation policy for ST_B first, and then extend the policy to that of ST_A.

3.3.1 Frame-Level Resource Allocation Policy for ST_B

Given a known P_P which is constant during one superframe, the optimal resource allocation policy for ST_B within a superframe is given by [27]

$$
\textbf{(P3-B)} \quad \max_{\mathbf{P}_{STBP} \in \mathscr{P}_{STBP}} \max_{\mathbf{P}_{STBS} \in \mathscr{P}_{STBS}} \left\{ \sum_{n=0}^{N-1} T_{WSum,B}(\mathbf{S}^{(n)}, P_P, P^{(n)}_{STBP}, P^{(n)}_{STBS}) \right\} \tag{3.16}
$$

where the CSI at the frame level evolves according to conditional probability $\Pr(\mathbf{S}^{(n+1)}|\mathbf{S}^{(n)})$ based on the fading profile. Denote the value function of problem P3-B within frames $\{n, n+1, \cdots, N-1\}$ as

$$
V_n(\mathbf{S}, P_P, X_{STB}) = \max_{\mathbf{P}_{STBP} \in \mathscr{P}_{STBP}} \max_{\mathbf{P}_{STBS} \in \mathscr{P}_{STBS}} \left\{ \sum_{k=n}^{N-1} T_{WSum,B}(\mathbf{S}^{(k)}, P_P, P_{STBP}^{(k)}, P_{STBS}^{(k)}) \right\}
$$

$$(3.17)$$

where $X_{STB}^n = X_{STB}$ and $\mathbf{S}^{(n)} = \mathbf{S}$. For $n \in [0, N-2]$, the dynamic programming equation of the value function is given by

$$
V_n(\mathbf{S}, P_P, X_{STB}) = \max_{\substack{P_{STBP}^{(n)} \in [0, P_{STBPM}^{(n)}], P_{STBS}^{(n)} \in [0, P_{STBM}] \\ P_{STBP}^{(n)} + P_{STBS}^{(n)} \leq \min\{P_{STBM}, \frac{X_{STB}}{(1-\beta)T}\}}} \left\{ T_{WSum,B}(\mathbf{S}^{(n)}, P_P, P_{STBP}^{(n)}, P_{STBS}^{(n)}) \right.
$$

$$
\left. + E\left[V_{n+1}(\mathbf{S}', P_P, X_{STB} - (1-\beta)T(P_{STBP}^{(n)} + P_{STBS}^{(n)}))|\mathbf{S}\right] \right\}
$$

$$(3.18)$$

where the conditional expectation is performed w.r.t. the conditional probability $\Pr(\mathbf{S}'|\mathbf{S})$ according to the fading profile. For $n = N-1$, we have

$$
V_{N-1}(\mathbf{S}, P_P, X_{STB}) = \max_{\substack{P_{STBP}^{N-1} \in [0, P_{STBPM}^{(N-1)}], P_{STBS}^{(N-1)} \in [0, P_{STBM}] \\ P_{STBP}^{(N-1)} + P_{STBS}^{(N-1)} \leq \min\{P_{STBM}, \frac{X_{STB}}{(1-\beta)T}\}}} \left\{ T_{WSum,B}(\mathbf{S}^{(N-1)}, P_P, P_{STBP}^{(N-1)}, P_{STBS}^{(N-1)}) \right\}.
$$

$$(3.19)$$

For notational simplicity, in the following analysis, we consider the nth frame as the tagged frame and denote $P_{STBP}^{(n)} = P_1$ and $P_{STBS}^{(n)} = P_2$, respectively. An illustration of the domain of (P_1, P_2) is given by Fig. 3.3. Define a resource allocation policy as follows:

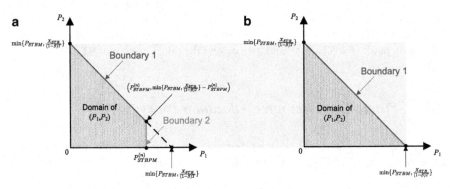

Fig. 3.3 Domain of (P_1, P_2): (a) $\min\{P_{STBM}, \frac{X_{STB}}{(1-\beta)T}\} \geq P_{STBPM}^{(n)}$; (b) $\min\{P_{STBM}, \frac{X_{STB}}{(1-\beta)T}\} < P_{STBPM}^{(n)}$

Definition 1 (Frame-Level Resource Allocation Policy of ST_B).

$$(\tilde{P}_1, \tilde{P}_2) = \begin{cases} (P_1^*, P_2^*), & \text{if } (P_1^*, P_2^*) \in \mathscr{D} \\ (P_{1B}^*, P_{2B}^*), & \text{otherwise} \end{cases} \tag{3.20}$$

where $\mathscr{D}$ is the domain of (P_1, P_2), as shown in Fig. 3.3. Denote $Q_n(\cdot)$ as the objective function to be optimized, i.e.,

$$Q_n(S, P_P, X_{STB}, P_1, P_2) = T_{WSum,B}(S, P_P, P_1, P_2) + E\left[V_{n+1}(S', P_P, X_{STB} - (1-\beta)T(P_1 + P_2))|S\right] \tag{3.21}$$

where $Q_{N-1}(S, P_P, X_{STB}, P_1, P_2) = T_{WSum,B}(S, P_P, P_1, P_2)$. Let $Q'(\cdot)$ and $Q''(\cdot)$ represent the two lines w.r.t. to boundary 1 and boundary 2, respectively, given by

$$Q'(S, P_P, X_{STB}, P_1) = Q_n\left(S, P_P, X_{STB}, P_1, \min\{P_{STBM}, \frac{X_{STB}}{(1-\beta)T}\} - P_1\right),$$

$$P_1 \in \left[0, \min\{P_{STBM}, \frac{X_{STB}}{(1-\beta_m)T}, P_{STBPM}^{(n)}\}\right] \tag{3.22}$$

$$Q''(S, P_P, X_{STB}, P_2) = Q_n\left(S, P_P, X_{STB}, P_{STBPM}^{(n)}, P_2\right), \quad P_2 \in \left[0, \min\{P_{STBM}, \frac{X_{STB}}{(1-\beta)T}\} - P_{STBPM}^{(n)}\right]. \tag{3.23}$$

Note that boundary 2 exists only when $\min\{P_{STBM}, \frac{X_{STB}}{(1-\beta)T}\} \geq P_{STBPM}^{(n)}$. Accordingly, in (3.20), (P_1^*, P_2^*) is an arbitrary point in the optimal set of $Q_n(\cdot)$ without considering the battery energy constraints, i.e., $(P_1^*, P_2^*) \in \{(P_1, P_2)|Q_n(S, P_P, X_{STB}, P_1, P_2) = Q^*\}$, where

$$Q^* = \max_{\substack{P_1 \in [0, P_{STBPM}^{(n)}], P_2 \in [0, P_{STBM}] \\ P_1 + P_2 \leq P_{STBM}}} \{Q_n(S, P_P, X_{STB}, P_1, P_2)\}. \tag{3.24}$$

Similarly, (P_{1B}^*, P_{2B}^*) is an arbitrary point in the optimal set of $Q_n(\cdot)$ on the boundary of the domain, given by [27]

$$(P_{1B}^*, P_{2B}^*) \in \{(P_1, P_2)|Q_n(S, P_P, X_{STB}, P_1, P_2) = Q_B^*,$$

$$P_1 \in \left[0, \min\{P_{STBM}, \frac{X_{STB}}{(1-\beta)T}, P_{STBPM}^{(n)}\}\right], P_2 = \min\{P_{STBM}, \frac{X_{STB}}{(1-\beta)T}\} - P_1,$$

$$\text{or } P_1 = P_{STBPM}^{(n)}, P_2 \in \left[0, \min\{P_{STBM}, \frac{X_{STB}}{(1-\beta)T}\} - P_{STBPM}^{(n)}\right]\} \tag{3.25}$$

where $Q_B^* = \max_{i \in \{1,2\}}\{Q_i^*\}$ with $Q_1^* = \max_{P_1 \in [0, \min\{P_{STBM}, \frac{X_{STB}}{(1-\beta)T}, P_{STBPM}^{(n)}\}]}\{Q'(S, P_P, X_{STB}, P_1)\}$ and $Q_2^* = \max_{P_2 \in [0, \min\{P_{STBM}, \frac{X_{STB}}{(1-\beta)T}\} - P_{STBPM}^{(n)}]}\{Q''(S, P_P, X_{STB}, P_2)\}$.

The rationale behind the policy is that, when the domain of (P_1, P_2) intersects with the optimal set, an arbitrary point in the intersection is chosen for resource allocation. Otherwise, an arbitrary point in the optimal set w.r.t. the boundary is chosen for resource allocation. In the following, we prove that the policy is optimal for problem P3-B. We begin with the following lemma for the weighted sum throughput w.r.t. ST_B:

Lemma 1. *The function* $T_{WSum,B}(\mathbf{S}, P_P, P_1, P_2)$ *is strictly concave w.r.t.* P_1 *and* P_2.

Proof 1. See Proof of Lemma 1 in Appendix.

For a general concave function, the following lemma holds:

Lemma 2. *For a concave function* $g(X)$ *and* $P_1 + P_2$ *in its domain. The function* $g(P_1 + P_2)$ *is concave w.r.t.* P_1 *and* P_2.

Proof 2. Consider two arbitrary points (P_1', P_2') and (P_1'', P_2'') with $P_1' + P_2'$ and $P_1'' + P_2''$ in the domain of $g(X)$. Based on the two points (P_1', P_2') and (P_1'', P_2''), consider an arbitrary line with the coordinates on the line given by $(P_1' + tP_1'', P_2' + tP_2'')$ and $P_1' + P_1'' + t(P_2' + P_2'')$ in the domain of $g(X)$. Since $P_1' + P_1'' + t(P_2' + P_2'')$ is an affine mapping of t and $g(t)$ is concave w.r.t. t, we have that $g(P_1' + P_1'' + t(P_2' + P_2''))$ is also concave. Since the necessary and sufficient condition for a function to be concave is that the function is concave when restricted to any line that intersects its domain [28], we have that $g(P_1 + P_2)$ is concave w.r.t. P_1 and P_2.

Then we show the concavity of the value function $V_n(\mathbf{S}, P_P, X)$ w.r.t. X. First consider three points w.r.t. to X, i.e., X_1, X_2, and X_3, where $X_1 \leq X_3$ and $X_2 = \theta X_1 + (1 - \theta)X_3$ with $0 \leq \theta \leq 1$. Denote the optimal points w.r.t. X_1 and X_3 as $(\tilde{P}_1(X_1), \tilde{P}_2(X_1))$ and $(\tilde{P}_1(X_3), \tilde{P}_2(X_3))$, respectively. We have the following lemma:

Lemma 3. *The point* $(\theta\tilde{P}_1(X_1) + (1-\theta)\tilde{P}_1(X_3), \theta\tilde{P}_2(X_1) + (1-\theta)\tilde{P}_2(X_3))$ *is in the domain defined w.r.t.* X_2, *i.e.,* $(\theta\tilde{P}_1(X_1) + (1-\theta)\tilde{P}_1(X_3), \theta\tilde{P}_2(X_1) + (1-\theta)\tilde{P}_2(X_3)) \in \mathscr{D}(X_2)$.

Proof 3. See Proof of Lemma 3 in Appendix.

For the effect of applying the threshold policy, we have the following lemma:

Lemma 4. *By applying the threshold policy on* $Q_n(\mathbf{S}, P_P, X, P_1, P_2)$, *the value function* $V_n(\mathbf{S}, P_P, X)$ *obtained after the operation is concave w.r.t.* X.

Proof 4. See Proof of Lemma 4 in Appendix.

Then we have the following theorem:

Theorem 1. *The resource allocation policy given by Definition 1 is optimal for problem P3-B.*

Proof 5. The proof is completed by induction. For $n = N - 1$, the optimality of the resource allocation policy is straightforward based on the concavity of $T_{WSum,B}(\cdot)$. Suppose the theorem holds for $n = k + 1$, based on Lemma 4, $V_{k+1}(\mathbf{S}, P_P, X)$ is concave w.r.t. X. It follows that $Q_k(\mathbf{S}, P_P, P_1, P_2)$ is concave w.r.t. P_1 and P_2 with the optimal policy for the kth frame being given by Definition 1. After applying the policy, $V_k(\mathbf{S}, P_P, X)$ is also concave w.r.t. X according to Lemma 4. This completes the proof.

Based on Definition 1, the values of Q^*, (P_1^*, P_2^*), Q_B^*, and (P_{1B}^*, P_{2B}^*) need to be calculated. The calculations of Q^* and (P_1^*, P_2^*) are straightforward since they are not dependent on the residual energy of ST_B. On the other hand, as the boundary of the domain of (P_1, P_2) is related to the residual energy of ST_B, the values of Q_B^* and (P_{1B}^*, P_{2B}^*) are functions of X_{STB}. However, by further investigating the problem, we can find simplified forms of the functions.

Let us consider boundary 2 first. Since the value function is concave w.r.t. P_1 and P_2, it is concave w.r.t. the restriction to any lines [28]. For boundary 2, the optimal value of $Q''(\cdot)$ without considering the energy and transmission power limitation is given by $\hat{Q}''^* = \max_{P_2 \geq 0}\{Q''(\mathbf{S}, P_P, X_{STB}, P_2)\}$, which can be achieved by an arbitrary optimal point $(\hat{P}_{1B2}^*, \hat{P}_{2B2}^*) \in \{(P_1, P_2)|Q_n(\mathbf{S}, P_P, X_{STB}, P_1, P_2) = \hat{Q}''^*\}$. For a concave function, since any local optimum is also global optimal and the optimal set is a convex set, the optimal point on boundary 2 can be calculated as

$$(P_{1B2}^*, P_{2B2}^*) = \begin{cases} (P_{STBPM}^{(n)}, \hat{P}_{2B2}^*), & \text{if } \hat{P}_{2B2}^* \in \left[0, \min\{P_{STBM}, \frac{X_{STB}}{(1-\beta)T}\} - P_{STBPM}^{(n)}\right] \\ (P_{STBPM}^{(n)}, \min\{P_{STBM}, \frac{X_{STB}}{(1-\beta)T}\} - P_{STBPM}^{(n)}), & \text{otherwise.} \end{cases}$$

$$(3.26)$$

Then we consider boundary 1. If $P_{STBM} \leq \frac{X_{STB}}{(1-\beta)T}$, since the line equation of boundary 1 is independent of X_{STB}, the optimal point (P_{1B1}^*, P_{2B1}^*) can be obtained in the same way as (3.26), given by

$$(P_{1B1}^*, P_{2B1}^*) = \begin{cases} (\hat{P}_{1B1}^*, \hat{P}_{2B1}^*), & \text{if } \hat{P}_{1B1}^* \in \left[0, P_{STBPM}^{(n)}\right] \\ (P_{STBPM}^{(n)}, \min\{P_{STBM}, \frac{X_{STB}}{(1-\beta)T}\} - P_{STBPM}^{(n)}), & \text{otherwise} \end{cases}$$

$$(3.27)$$

where $\hat{Q}'^* = \max_{P_1 \geq 0}\{Q'(\mathbf{S}, P_P, X_{STB}, P_1)\}$ and $(\hat{P}_{1B1}^*, \hat{P}_{2B1}^*) \in \{(P_1, P_2)|Q_n(\mathbf{S}, P_P, X_{STB}, P_1, P_2) = \hat{Q}'^*\}$. On the other hand, if $P_{STBM} > \frac{X_{STB}}{(1-\beta)T}$, since the resource allocation on this boundary saturates the residual energy, there is no energy for further communications in the subsequent frames. Therefore, the optimal point on boundary 1 is obtained by maximizing the first term of $Q_n(\cdot)$, i.e.,

$$(\hat{P}_{1B1}^*, \hat{P}_{2B1}^*) \in \left\{(P_1, \frac{X_{STB}}{(1-\beta)T} - P_1)|T_{WSum,B}(\mathbf{S}, P_P, P_1, \frac{X_{STB}}{(1-\beta)T} - P_1) = \hat{Q}'^*\right\}$$

$$(3.28)$$

where $\hat{Q}'^* = \max_{P_1 \geq 0}\{T_{WSum,B}(\mathbf{S}, P_P, P_1, \frac{X_{STB}}{(1-\beta)T} - P_1)\}$. According to Lemma 1, since $T_{WSum,B}(\mathbf{S}, P_P, P_1, P_2)$ is strictly concave w.r.t. P_1 and P_2, it is also strictly concave by restricting to any lines. Therefore, $T_{WSum,B}(\mathbf{S}, P_P, P_1, \frac{X_{STB}}{(1-\beta)T} - P_1)$ has a single stationary point w.r.t. P_1. Let the first order derivative of $T_{WSum,B}(\mathbf{S}, P_P, P_1, \frac{X_{STB}}{(1-\beta)T} - P_1)$ equal to 0, we can obtain the closed-form expression of $\hat{P}_{1B1}^*$ w.r.t. X_{STB}, given by

$$\hat{P}^*_{1B1} = \frac{(\delta_1 + \delta_2)N_0 + \delta_1 |h^{(n)}_{BB}|^2 \frac{X_{STB}}{(1-\beta)T} + \delta_2 |h^{(n)}_{AP}| P_{SRA}}{\delta_1 |h^{(n)}_{BB}|^2 - \delta_2 |h^{(n)}_{BP}|^2} \tag{3.29}$$

where $\delta_1 = \frac{(1-\zeta)(1-\beta)T|h^{(n)}_{BP}|^2}{N_0}$ and $\delta_2 = -\frac{\zeta(1-\beta)T|h^{(n)}_{BB}|^2}{N_0}$. Then, the optimal value of P_2 on boundary 1 is given by $\hat{P}^*_{2B1} = \min\{P_{STBM}, \frac{X_{STB}}{(1-\beta)T}\} - \hat{P}^*_{1B1}$, and $\hat{Q}'^*$ can be obtained using (3.21).

3.3.2 Frame-Level Resource Allocation Policy for ST_A

The objective of the resource allocation is to maximize the data volume delivery from ST_A to SR_A given the polarization characteristic γ, the initial state of fading characteristic $h^{(0)}$, and the initial residual energy $X^{(0)}_{STA}$. Denote the transmission power of ST_A within the nth frame as $P^{(n)}_{STA}$. The resource allocation problem of ST_A is given by

$$\textbf{(P3-A)} \quad \max_{P_{STA} \in \mathscr{P}_{STA}} \left\{ \sum_{n=0}^{N-1} T_{WSum,A}(\mathbf{S}, X^{(n)}_{STA}, P^{(n)}_{STA}) \right\}. \tag{3.30}$$

The resource allocation for ST_A can be considered as a special case of that for ST_B without the power allocation for relaying the primary user's packets. Hence, the domain of $P^{(n)}_{STA}$ is a line segment instead of an area. Therefore, the optimal resource allocation policy is given by Definition 2 below. We omit the proof of the optimality in this chapter because of space limitation.

Definition 2 (Frame-Level Resource Allocation Policy for ST_A).

$$\tilde{P}_1 = \begin{cases} P^*_1, & \text{if } P^*_1 \in \left[0, \min\left\{P_{STAM}, \frac{X_{STA}}{\beta T}\right\}\right] \\ \min\left\{P_{STAM}, \frac{X_{STA}}{\beta T}\right\}, & \text{otherwise} \end{cases} \tag{3.31}$$

where $P^*_1 \in \arg\max_{P_1 \in [0, P_{STAM}]} \tilde{Q}_n(\mathbf{S}, X_{STA}, P_1)$, and $\tilde{Q}_n(\cdot)$ is given by

$$\tilde{Q}_n(\mathbf{S}, X_{STA}, P_1) = T_{WSum,A}(\mathbf{S}, P_1) + E\left[\tilde{V}_{n+1}(\mathbf{S}', X_{STA} - \beta T P_1)|\mathbf{S}\right] \tag{3.32}$$

where $\tilde{Q}_{N-1}(\mathbf{S}, X_{STA}, P_1) = T_{WSum,A}(\mathbf{S}, P_1)$. For $n \in [0, N-2]$,

$$\tilde{V}_n(\mathbf{S}, X_{STA}) = \max_{P^{(n)}_{STA} \in \left[0, \min\left\{P_{STAM}, \frac{X_{STA}}{\beta T}\right\}\right]} \left\{T_{WSum,A}(\mathbf{S}, P^{(n)}_{STA}) + E\left[\tilde{V}_{n+1}(\mathbf{S}', X_{STA} - \beta T P^{(n)}_{STA})|\mathbf{S}\right]\right\}. \tag{3.33}$$

For $n = N - 1$, we have

$$\tilde{V}_{N-1}(\mathbf{S}, X_{STA}) = \max_{P_{STA}^{(N-1)} \in \left[0, \min\left\{P_{STAM}, \frac{X_{STA}}{\beta T}\right\}\right]} \left\{T_{WSum,A}(\mathbf{S}, P_{STA}^{(N-1)})\right\}. \tag{3.34}$$

3.3.3 Superframe-Level Resource Allocation Policy

Unlike the resource allocation policies for SUs in the frame level, the resource allocation for the PU cannot be analyzed by using the above analytical method, i.e., it is impossible to get a close-form solution. Here, we propose to use the heuristic online methods in [26] to get the superframe-level resource allocation policy for the PU. The parallel rollout scheme which is based on the decision rule/policy improvement principle in the policy iteration algorithm in [29] can be adopted here. Since the value of any decision rule pair is a lower bound to the optimal value, the parallel rollout iteration is a lower bound scheme. The upper bound based iteration can also be used for the PU, for example, hindsight approach in [30]. The number of iterations can be determined by many factors such as the time required to obtain the optimal solution. If the parallel rollout is used for the PU, the lower bound of the iteration approach is given by $|Q^*(\mathbf{S}, P_P) - \hat{Q}(\mathbf{S}, P_P)| \leq \frac{2\eta\epsilon}{1-\eta}$, where $Q^*(\mathbf{S}, P_P) = \max_{P_P}\{T_{WSum}\}$ is the analytically optimized value of the PU, $\hat{Q}(\mathbf{S}, P_P)$ is the iteration solution, ϵ is defined as $\sup_{P_P} |Q^*(\mathbf{S}, P_P) - U(\mathbf{S}, P_P)| \leq \epsilon$, and $U(\mathbf{S}, P_P)$ is a bounded and measurable function which is determined by iteration times. Due to space limitations, the detailed iteration algorithms are omitted.

3.4 Approximation Algorithm and Numerical Results

To simplify the calculation and interpretation of the proposed policies, two modified backward iteration based approximation algorithms for ST_A and ST_B policies are given in Algorithm 1 and Algorithm 2, respectively. For simplicity of optimization, the calculation of Shannon capacity in the iteration process is approximated by linear fitting. The power policy obtained by the approximation backward iteration is used to calculate the real reward functions without approximation. Note that step 10 in Algorithm 1 and step 20 in Algorithm 2 involve two convex optimization problems which can be readily solved by sophisticated algorithms [28], according to the stochastic inventory theory [31, 32].

Since the channel is independent with each other in both frame and superframe level, the problem is solved by decomposing it into two independent subproblems, i.e., ST_A policy and ST_B policy. Therefore, the simulations are done using Matlab for ST_A and ST_B separately. Channel models with Markovian properties are adopted in our simulations. Two scenarios are considered in our simulations. Specifically,

---

Algorithm 1 Backward Iteration Algorithm of ST_A

Input: maximum batter capacity X_{STAM}, initial energy level X_{STA}, maximum transmit power P_{STAM}, channel state set $\{h_{AA}\}$, transition probability matrix of channel state, steady-state probability of each state, noise power N_0, β, ζ, η, T, number of frames in one superframe N,

Output: Optimal $P_{STA}^{(n)}$;

1: **if** $n = N$ **then**
2:      **if** $X_{STA}^{N} \geq P_{STAM}\beta T$ **then**
3:          $P_{STA}^{(N)} = P_{STAM}$;
4:      **else**
5:          $P_{STA}^{(N)} = \max\left(0, \frac{X_{STA}^{(N)}}{\beta T}\right)$;
6:      **end if**
7: **end if**
8: **for** $n = N - 1$ to 1 **do**
9:      **for** Channel State Transition $h_{AA}^{(n)}(i) \to h_{AA}^{(n+1)}(j)$ **do**
10:         Calculate optimal $P_{STA}^{(n)}$ according to (3.33);
11:     **end for**
12: **end for**
13: **return**

---

in the first scenario, each user is uniformly distributed in the area of 1000×1000 m^2 with low speed, e.g., $v = \{1, 2\}$ m/s. To investigate the performance in the high speed environment, we simply model the channel state as a two state on-off channel in the second scenario, as discussed in [33] where the velocity is up to 30 m/s. In the first scenario, each channel has nine states in terms of SNR listed as SNR $= \{1, 6.02, 7.78, 9.03, 10.79, 17.04, 18.80, 24.05, 24.56\}$ in dB, and the path loss exponent is 3. In the second scenario, the *off* state is with a SNR of 0 dB and the *on* state is with a SNR of 14.77 dB, respectively.

Time duration of each frame is $T = 100$ ms, and the weighting parameter and the discount factor are $\zeta = 0.4$ and $\eta = 1$, respectively. The bandwidth in our system is 1 MHz. The associated channel state transition probability matrix and the steady-state probability of on-off and 9-state channels can be obtained by using these parameters based on the Markov model described in [33] and [35], respectively. The reference path loss and the reference distance in our simulations are 5.105 and 100 m.

In the simulation of ST_A power policy, we consider both on-off Markov channel and 9-state Markov channel. The maximum transmit power is $P_{STAM} = 5 \times 10^{-3}$ W, and the initial energy of S_{TA} is $X_{STA} = 1.2 \times 10^{-3}$ W s. Note that the energy consumption of one frame using P_{STAM} is $P_{STAM}\beta T = 2.5 \times 10^{-4}$ W s. We compare the result obtained using Algorithm 1 with the greedy power policy and the random power policy. In the greedy power policy, ST_A always maximizes the throughput of each frame, i.e., transmits at its maximum transmit power. In the random power policy, ST_A randomly chooses a transmit power not greater than P_{STAM} at each frame. The throughput performance of each power policy under different conditions is shown in Fig. 3.4.

Algorithm 2 Backward Iteration Algorithm of ST_B

Input: maximum batter capacity X_{STBM}, initial energy level X_{STB}, maximum transmit power P_{STBM}, channel state sets $\{h_{BB}\}, \{h_{PA}\}, \{h_{PB}\}, \{h_{BP}\}$, transition probability matrix of each channel, steady-state probability matrix of each channel, noise power N_0, β, ζ, η, T, number of frames in one superframe N,

Output: Optimal $P_{STBP}^{(n)}, P_{STBS}^{(n)}$;

1: **if** $n = N$ **then**
2:     **if** $X_{STB}^{(N)} \geq P_{STBM}(1 - \beta)T$ **then**
3:        **if** $T_P \geq T_{SB}$ in (3.15) **then**
4:           $P_{STBP}^{(n)} = \min\left(P_{STBM}, P_{STBPM'}^{(N)}\right)$ and $P_{STBS}^{(n)} = P_{STBM} - P_{STBP}^{(n)}$;
5:        **else**
6:           $P_{STBP}^{(n)} = 0$ and $P_{STBS}^{(n)} = P_{STBM}$;
7:        **end if**
8:     **else**
9:        **if** $T_P \geq T_{SB}$ in (3.15) **then**
10:          $P_{STBP}^{(n)} = \min\left(\frac{X_{STB}^{(N)}}{(1-\beta)T}, P_{STBPM'}^{(N)}\right)$ and $P_{STBS}^{(n)} = \frac{X_{STB}^{(N)}}{(1-\beta)T} - P_{STBP}^{(n)}$;
11:       **else**
12:          $P_{STBP}^{(n)} = 0$ and $P_{STBS}^{(n)} = \max\left(0, \frac{X_{STB}^{(N)}}{(1-\beta)T}\right)$;
13:       **end if**
14:    **end if**
15: **end if**
16: **for** $n = N - 1$ to 1 **do**
17:    **for** Channel State Transition $h_{BB}^{(n)}(i) \rightarrow h_{BB}^{(n+1)}(j)$ **do**
18:       **for** Channel State Transition $h_{BP}^{(n)}(i) \rightarrow h_{BP}^{(n+1)}(j)$ **do**
19:          **for** Channel State Transition $h_{PB}^{(n)}(i) \rightarrow h_{PB}^{(n+1)}(j)$ **do**
20:             Calculate optimal $P_{STBP}^{(n)}$ and $P_{STBS}^{(n)}$ according to (3.18);
21:          **end for**
22:       **end for**
23:    **end for**
24: **end for**
25: **return**

Specifically, $T_{WSum,A}$ of different frame numbers in one superframe under the on-off Markov channel is shown in Fig. 3.4a. Since X_{STA} is only enough for less than $N_{STAM} = 5$ times transmission at the power level of P_{STAM}, it can be seen that MDP based policy has the same throughput with that of greedy policy when $N = 4$. This shows MDP policy can achieve the same performance when the energy is enough for one superframe transmission. We can find that the throughput of MDP based power policy increases with the increasing of N. The rationale behind this result is that with more number of frames, the probability that ST_A transmits at the good channel condition increases. The MDP policy in this two state on-off model can be interpreted as ST_A aims to transmit as much as possible when the channel is good, while ST_A tries to save energy for future good channels when the current channel is bad. In this regard, the optimal performance for MDP happens when ST_A transmits using P_{STAM} only when the channel is good, while keeping idle when the channel is bad. This optimal performance requires a sufficiently large N to enable

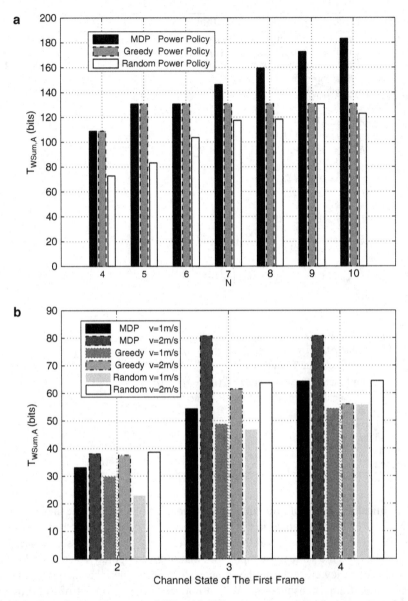

Fig. 3.4 Throughput performance of ST_A with different number of frames in one superframe, different initial channel states and velocities under different Markov channel models and power policies. (**a**) ST_A's throughput comparison among MDP, greedy, and random power policies with different number of frames under the on-off channel model. (**b**) ST_A's throughput comparison among MDP, greedy, and random power policies with different initial channel states and velocities under the 9-state channel model ($N = 7$)

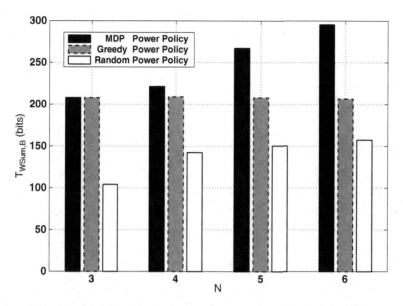

Fig. 3.5 Performance comparison using MDP, greedy, and random power policies with different number of frames under the on-off Markov channel model for ST_B policy

ST_A to consume its energy only in the on state. It is evident that the throughput of greedy based power policy is a constant after $N = 5$, as X_{STA} can only guarantee five times transmissions. The throughput using random power policy increases with increasing N, because the probability that ST_A can transmit in the good channel condition increases when N is large. It is noted that the MDP and greedy policies achieve the same throughput when N is 5 and 6 as shown in Fig. 3.4a. The reason for this result comes from that channel conditions for transmissions using MDP are most likely the same as that of using the greedy policy, as the maximum number of frames in one superframe is near N_{STAM}.

Performance comparison of ST_A using MDP, greedy and random power policies in the 9-state Markov channel model with different velocities and initial channel states is shown in Fig. 3.4b. For a given velocity, it is straightforward that all these three policies achieve better performance when the initial channel state, i.e., the channel state of the first frame, is under a better SNR condition, as shown in the figure. When the initial channel condition is not sufficiently good, for a given initial channel state, it can be seen from the figure that the throughput increases with the increasing of velocity. Since the channel state in current frame will more likely to transit to the next channel state. In this context, ST_A can transit from a low SNR to a high one with a higher probability at a higher velocity. It can be seen from that the proposed MDP power policy outperforms both greedy and random power policies with different initial channel states and velocities under the 9-state channel model.

To validate the effectiveness of the ST_B policy, performance comparison of $T_{WSum,B}$ using MDP, greedy, and random power policies is investigated. The

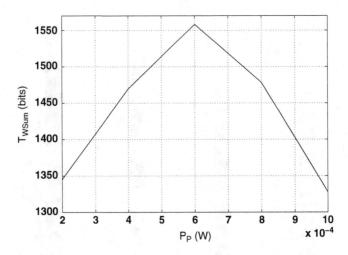

Fig. 3.6 Relationship between T_{WSum} of one superframe and the optimal PU's transmit power

maximum transmit power is $P_{STBM} = 8 \times 10^{-3}$ W, and the initial energy of ST_B is $X_{STB} = 1.2 \times 10^{-3}$ W s. Note that the energy consumption of one frame using P_{STBM} is $P_{STBM}(1 - \beta)T = 4 \times 10^{-4}$ W s. The transmit power of the PU is $P_P = 2 \times 10^{-3}$ W. The optimal MDP policy using Algorithm 2 for ST_B is that ST_B aims to transmit on the channel with higher throughput at each frame, e.g., when $T_P^{(n)} \geq T_{SB}^{(n)}$ in (3.15), ST_B will forward the PU's data using P_{STBPM} in the nth frame, otherwise ST_B transmits its own data with the power level of P_{STBM}. In this simulation, all of $\{h_{BB}\}, \{h_{PA}\}, \{h_{PB}\}, \{h_{BP}\}$ are modeled based on the on-off Markov channel models for computational simplicity. It can be seen from Fig. 3.5 that MDP based power policy can get higher throughput with an increasing N. Since X_{STB} can only guarantee three times transmissions when the greedy power policy is used, the throughput using greedy policy is a constant for $N \geq 3$. The proposed MDP policy of ST_B outperforms both greedy and random power policies.

To investigate the superframe level policy, numerical result of optimal P_P is given in Fig. 3.6. XPD of each superframe is assumed to be independent and identically distributed [34]. Since spatial channel fading between two superframes is also independent and identically distributed, the PU can transmit using a constant power for all superframes. In this context, the optimization problem in the superframe level is to find the optimal P_P to reach the maximum T_{WSum}. The initial energy level of the PU is $X_P = 1 \times 10^{-3}$ W s. The channel model of spatial fading in each superframe for ST_A is according to the on-off model, and that of ST_B is the 9-state SNR model with $v = 1$ ms, respectively. In each superframe, there are $N = 7$ frames, and $M = 4$ superframes for the PU are considered. It can be seen that optimal P_P that can reach the maximum T_{WSum} is achieved at 10^{-4} W. The rationale behind this result is that, with a low P_P, although the PU still has energy after four superframes cooperation, the PU contributes little to the weighted sum throughput of each superframe. With the increasing of P_P, although the weighted sum throughput of each superframe increases, the PU depletes its energy which terminate the cooperative communications between the PU and SUs in later superframes. Therefore, we can conclude that the optimal value of P_P is 6×10^{-4}. To further reduce the complexity of computing the optimal PP, the heuristic online method can be used, and the performance bound is given mathematically.

3.5 Conclusions

In this chapter, we have proposed a novel polarization enabled two-phase cooperation framework for CCRN. By utilizing ODPAs at SU transceivers, SUs can simultaneously relay the PU's data and transmit their own information without mutual interference. We have modeled this system as a two time-scales cooperation framework by taking both the spatial and polarization domains into consideration. To evaluate the effectiveness of the proposed framework, a sum throughput maximization problem with throughput and power constraints has been formulated. The optimization problem has been analyzed and solved by using MDP. In practice, CSI is imperfect; our future work will address the problem of imperfect CSI.

Appendix

Proof of Lemma 1. Since $T_{WSum,B}(\mathbf{S}, P_P, P_1, P_2)$ is twice continuously differentiable w.r.t. P_1 and P_2, the proof is completed by investigating the Hessian $\mathbf{H} = [H_{ii}]_{2\times2}$ of $T_{WSum,B}(\mathbf{S}, P_P, P_1, P_2)$. We have

$$H_{11} = \frac{\partial^2 T_{WSum,B}(\mathbf{S}, P_P, P_1, P_2)}{\partial P_1^2} = -\frac{(1-\zeta)(1-\beta)TG^2|h_{BP}^{(n)}|^4}{\left(N_0 + Q|h_{AP}^{(n)}|^2 P_{SRA}^{(n)} + G|h_{BP}^{(n)}|^2 P_1\right)^2} \tag{3.35}$$

$$H_{22} = \frac{\partial^2 T_{WSum,B}(\mathbf{S}, P_P, P_1, P_2)}{\partial P_2^2} = -\frac{\zeta(1-\beta)T|h_{BB}^{(n)}|^4}{\left(N_0 + |h_{BB}^{(n)}|^2 P_2\right)^2} \tag{3.36}$$

$$H_{12} = \frac{\partial^2 T_{WSum,B}(\mathbf{S}, P_P, P_1, P_2)}{\partial P_1 \partial P_2} = 0 \tag{3.37}$$

$$H_{21} = \frac{\partial^2 T_{WSum,B}(\mathbf{S}, P_P, P_1, P_2)}{\partial P_2 \partial P_1} = 0. \tag{3.38}$$

Since all off-diagonal elements of $\mathbf{H}$ equal to zero, H_{11} and H_{22} are the two eigenvalues of $\mathbf{H}$. Since $\zeta \in (0, 1)$ and $\beta \in (0, 1)$, we have $H_{11} < 0$ and $H_{22} < 0$. Therefore, $\mathbf{H}$ is negative definite, which indicates that $T_{WSum,B}(\mathbf{S}, P_P, P_{STBP}^{(n)}, P_{STBS}^{(n)})$ is strictly concave w.r.t. to P_1 and P_2.

Proof of Lemma 3. Consider boundary 2 of $\mathscr{D}(X_1)$ and $\mathscr{D}(X_3)$, we have $\tilde{P}_1(X_1), \tilde{P}_1(X_3) \leq P_{STBPM}^{(n)}$. Therefore, we can easily verify that $\theta\tilde{P}_1(X_1) + (1 - \theta)\tilde{P}_1(X_3) \leq P_{STBPM}^{(n)}$. In other words, $(\theta\tilde{P}_1(X_1) + (1 - \theta)\tilde{P}_1(X_3), \theta\tilde{P}_2(X_1) + (1 - \theta)\tilde{P}_2(X_3))$ is on boundary 2 of $\mathscr{D}(X_2)$.

Then we investigate boundary 1 of $\mathscr{D}(X_2)$. Since the upper bound of $P_1 + P_2$ is $\min\{P_{STBM}, \frac{X_{STB}}{(1-\beta)T}\}$, which depends on the value of a constant P_{STBM}, we consider the four cases in the proof: Case 1: $\frac{X_3}{(1-\beta)T} < P_{STBM}$; Case 2: $\frac{X_2}{(1-\beta)T} < P_{STBM} \leq \frac{X_3}{(1-\beta)T}$; Case 3: $\frac{X_1}{(1-\beta)T} < P_{STBM} \leq \frac{X_2}{(1-\beta)T}$; Case 4: $P_{STBM} \leq \frac{X_1}{(1-\beta)T}$. All other cases are not considered since they contradict with the fact that $X_1 \leq X_2 \leq X_3$. The proofs of all cases are given as follows:

Case 1:   We have

$$\tilde{P}_1(X_1) + \tilde{P}_2(X_1) \leq \min\left\{P_{STBM}, \frac{X_1}{(1-\beta)T}\right\} = \frac{X_1}{(1-\beta)T} \tag{3.39}$$

$$\tilde{P}_1(X_3) + \tilde{P}_2(X_3) \leq \min\left\{P_{STBM}, \frac{X_3}{(1-\beta)T}\right\} = \frac{X_3}{(1-\beta)T}. \tag{3.40}$$

Then, for the point under consideration, we have

$$\theta\tilde{P}_1(X_1) + (1 - \theta)\tilde{P}_1(X_3) + \theta\tilde{P}_2(X_1) + (1 - \theta)\tilde{P}_2(X_3) \leq \theta\frac{X_1}{(1-\beta)T} + (1 - \theta)\frac{X_3}{(1-\beta)T}$$

$$= \frac{X_2}{(1-\beta)T} = \min\left\{P_{STBM}, \frac{X_2}{(1-\beta)T}\right\}. \tag{3.41}$$

Case 2:   We have

$$\tilde{P}_1(X_1) + \tilde{P}_2(X_1) \leq \min\left\{P_{STBM}, \frac{X_1}{(1-\beta)T}\right\} = \frac{X_1}{(1-\beta)T} \tag{3.42}$$

$$\tilde{P}_1(X_3) + \tilde{P}_2(X_3) \leq \min\left\{P_{STBM}, \frac{X_3}{(1-\beta)T}\right\} = P_{STBM}. \tag{3.43}$$

Then, for the point under consideration, we have

$$\theta\tilde{P}_1(X_1) + (1 - \theta)\tilde{P}_1(X_3) + \theta\tilde{P}_2(X_1) + (1 - \theta)\tilde{P}_2(X_3) \leq \theta\frac{X_1}{(1-\beta)T} + (1 - \theta)P_{STBM}$$

$$\leq \theta\frac{X_1}{(1-\beta)T} + (1 - \theta)\frac{X_3}{(1-\beta)T} = \frac{X_2}{(1-\beta)T} = \min\left\{P_{STBM}, \frac{X_2}{(1-\beta)T}\right\}. \tag{3.44}$$

Case 3:   We have

$$\tilde{P}_1(X_1) + \tilde{P}_2(X_1) \leq \min\left\{P_{STBM}, \frac{X_1}{(1-\beta)T}\right\} = \frac{X_1}{(1-\beta)T} \qquad (3.45)$$

$$\tilde{P}_1(X_3) + \tilde{P}_2(X_3) \leq \min\left\{P_{STBM}, \frac{X_3}{(1-\beta)T}\right\} = P_{STBM}. \qquad (3.46)$$

Then, for the point under consideration, we have

$$\theta\tilde{P}_1(X_1) + (1-\theta)\tilde{P}_1(X_3) + \theta\tilde{P}_2(X_1) + (1-\theta)\tilde{P}_2(X_3) \leq \theta\frac{X_1}{(1-\beta)T} + (1-\theta)P_{STBM}$$

$$\leq \theta P_{STBM} + (1-\theta)P_{STBM} = P_{STBM} = \min\left\{P_{STBM}, \frac{X_2}{(1-\beta)T}\right\}. \qquad (3.47)$$

Case 4:   We have

$$\tilde{P}_1(X_1) + \tilde{P}_2(X_1) \leq \min\left\{P_{STBM}, \frac{X_1}{(1-\beta)T}\right\} = P_{STBM} \qquad (3.48)$$

$$\tilde{P}_1(X_3) + \tilde{P}_2(X_3) \leq \min\left\{P_{STBM}, \frac{X_3}{(1-\beta)T}\right\} = P_{STBM}. \qquad (3.49)$$

Then, for the point under consideration, we have

$$\theta\tilde{P}_1(X_1) + (1-\theta)\tilde{P}_1(X_3) + \theta\tilde{P}_2(X_1) + (1-\theta)\tilde{P}_2(X_3) \leq \theta P_{STBM} + (1-\theta)P_{STBM}$$

$$= P_{STBM} = \min\left\{P_{STBM}, \frac{X_2}{(1-\beta)T}\right\}. \qquad (3.50)$$

In other words, $(\theta\tilde{P}_1(X_1) + (1-\theta)\tilde{P}_1(X_3), \theta\tilde{P}_2(X_1) + (1-\theta)\tilde{P}_2(X_3))$ is also on boundary 1 of $\mathscr{D}(X_2)$, which completes the proof.

Proof of Lemma 4. We first show that the resource allocation policy given by Definition 1 can achieve the optimal value of $Q_n(S, P_P, X, P_1, P_2)$, given $Q_n(S, P_P, X, P_1, P_2)$ is concave w.r.t. (P_1, P_2). Since boundary 1 is related to the value of X which complicates the analysis, we first investigate a "loose" version of boundary 1 without considering the residual battery energy, i.e., $P_1 + P_2 = P_{STBM}$. Combined with boundary 2, we define an auxiliary domain $\mathscr{D}'$ which is independent of X. Then, the optimal value of $Q_n(S, P_P, X, P_1, P_2)$ in $\mathscr{D}'$ is given by (3.24) at point (P_1^*, P_2^*). Then, we investigate the original domain $\mathscr{D}(X)$. If the point (P_1^*, P_2^*) is in $\mathscr{D}(X)$, then the optimal value of $Q_n(S, P_P, X, P_1, P_2)$ (i.e., $V_n(S, P_P, X)$) is given by Q^*. On the other hand, if (P_1^*, P_2^*) is not in $\mathscr{D}(X)$, the optimal value of $Q_n(S, P_P, X, P_1, P_2)$ should be achieved on the boundary (either boundary 1 or boundary 2) of $\mathscr{D}(X)$. The reason is that, $Q_n(S, P_P, X, P_1, P_2)$ is concave w.r.t. P_1 and P_2, which implies that any local optimum is also the global optimum. Moreover, since the optimal set of a convex function is also convex, there is no isolated optimal set within $\mathscr{D}(X)$. Therefore, we only need to calculate the optimal value of $Q_n(\mathbf{S}, P_P, X, P_1, P_2)$ on the boundary. The optimal point is given by (P_{1B}^*, P_{2B}^*) in Definition 1. Therefore, the resource allocation policy achieves the optimal value of $Q_n(\mathbf{S}, P_P, X, P_1, P_2)$, given it is concave.

Then, we show the concavity of $V_n(\mathbf{S}, P_P, X)$ for any n. The proof is completed by induction. For $n = N - 1$, the concavity of $Q_{N-1}(\mathbf{S}, P_P, X, P_1, P_2)$ (defined according to (3.21) for the $(N - 1)$th frame) w.r.t. P_1 and P_2 is straightforward based on the concavity of function $T_{WSum,B}(\cdot)$. Based on the concavity of $Q_{N-1}(\cdot)$ and the definitions of points $(\tilde{P}_1(X_1), \tilde{P}_2(X_1))$, $(\tilde{P}_1(X_3), \tilde{P}_2(X_3))$, and $(\theta\tilde{P}_1(X_1) + (1 - \theta)\tilde{P}_1(X_3), \theta\tilde{P}_2(X_1) + (1 - \theta)\tilde{P}_2(X_3))$ according to Lemma 3, we have

$$\theta V_{N-1}(\mathbf{S}, P_P, X_1) + (1 - \theta)V_{N-1}(\mathbf{S}, P_P, X_3)$$

$$= \theta Q_{N-1}(\mathbf{S}, P_P, X_1, \tilde{P}_1(X_1), \tilde{P}_2(X_1)) + (1 - \theta)Q_{N-1}(\mathbf{S}, P_P, X_3, \tilde{P}_1(X_3), \tilde{P}_2(X_3))$$

$$\leq Q_{N-1}(\mathbf{S}, P_P, \theta X_1 + (1 - \theta)X_3, \theta\tilde{P}_1(X_1) + (1 - \theta)\tilde{P}_1(X_3), \theta\tilde{P}_2(X_1) + (1 - \theta)\tilde{P}_2(X_3)))$$

$$= Q_{N-1}(\mathbf{S}, P_P, X_2, \theta\tilde{P}_1(X_1) + (1 - \theta)\tilde{P}_1(X_3), \theta\tilde{P}_2(X_1) + (1 - \theta)\tilde{P}_2(X_3)))$$

$$\leq Q_{N-1}(\mathbf{S}, P_P, X_2, \tilde{P}_1(X_2), \tilde{P}_2(X_2))$$

$$= V_{N-1}(\mathbf{S}, P_P, X_2) \tag{3.51}$$

where the last inequality holds because the point $(\theta\tilde{P}_1(X_1) + (1 - \theta)\tilde{P}_1(X_3), \theta\tilde{P}_2(X_1) + (1 - \theta)\tilde{P}_2(X_3))$ lies in $\mathscr{D}(X_2)$ according to Lemma 3, within which the optimal value is given by $Q_{N-1}(\mathbf{S}, P_P, \tilde{P}_1(X_2), \tilde{P}_2(X_2))$. Based on the definition of X_1, X_2, and X_3 and the definition of concavity, we have that $V_{N-1}(\mathbf{S}, P_P, X)$ is also concave w.r.t. X. Then, suppose the lemma holds for $n = k + 1$, i.e., $V_{k+1}(\mathbf{S}, P_P, X)$ is concave w.r.t. X. According to the theory of convex optimization[28], since expectation can be considered as a nonnegative weighted sum, $E[V_{k+1}(\mathbf{S}', P_P, X)|\mathbf{S}]$ is also concave w.r.t. to X. Following the same steps as (3.51), we can prove that $V_k(\mathbf{S}, P_P, X)$ is concave w.r.t. X, which completes the proof.

References

1. FCC, "Spectrum policy task force," *ET Docket 02-135*, Nov. 2002.
2. I. Akyildiz *et al.* "Next generation/dynamic spectrum access/cognitive radio wireless networks: A survey," *Comput. Netw.*, vol. 50, pp. 2127–2159, May 2006
3. F. Granelli, P. Pawelczak, R. Prasad, K. Subbalakshmi, R. Chandramouli, J. Hoffmeyer, and H. Berger, "Standardization and research in cognitive and dynamic spectrum access networks: IEEE SCC41 efforts and other activities," *IEEE Commun. Mag.*, vol. 48, no. 1, pp. 71–79, Jan. 2010.
4. J. Wang, M. Ghosh, and K. Challapali, "Emerging cognitive radio applications: A survey," *IEEE Commun. Mag.*, vol. 49, no. 3, pp. 74–81, Mar. 2011.

5. B. Wang, and K. Liu, "Advances in cognitive radio networks: A survey," *IEEE J. Sel. Top. Sig. Proc.*, vol. 5, no. 1, pp. 5–23, Feb. 2011.
6. Y. Liang, K. Chen, G. Li, and P. Mahonen, "Cognitive radio networking and communications: An overview," *IEEE Trans. on Veh. Tech.*, vol. 60, no. 7, pp. 3386–3407, Sept. 2011.
7. A. Goldsmith, S. A. Jafar, I. Maric, and S. Srinivasa, "Breaking spectrum gridlock with cognitive radios: An information theoretic perspective," *Proc. IEEE*, vol. 97, no. 5, pp. 894–914, May, 2009.
8. O. Simeone, I. Stanojev, S. Savazzi, Y. Bar-Ness, U. Spagnolini, and R. Pickholtz. "Spectrum leasing to cooperating secondary ad hoc networks", *IEEE J. Sel. Areas Commun.*, vol. 26, no. 1, pp. 203–213, Jan. 2008.
9. J. Zhang and Q. Zhang, "Stackelberg game for utility-based cooperative cognitive radio networks," in *Proc. MobiHoc*, New Orleans, LA, May 2009.
10. H. Xu and B. Li, "Efficient resource allocation with flexible channel cooperation in OFDMA cognitive radio networks," in *Proc. IEEE INFOCOM*, San Diego, CA, Mar. 2010.
11. W. Su, J. D. Matyjas, and S. Batalama, "Active cooperation between primary users and cognitive radio users in cognitive ad-hoc networks," *IEEE Trans. Signal Proc.*, vol. 60, no. 4, pp. 1796–1805, Apr. 2012.
12. B. Cao, J. W. Mark, Q. Zhang, R. Lu, X. Lin, and S. Shen, "On optimal communication strategies for cooperative cognitive radio networking," in *Proc. INFOCOM*, Turin, Italy, Apr. 2013.
13. B. Cao, Y. Cui, Q. Zhang, and J. W. Mark, "Game theoretic analysis of orthogonal modulation based cooperative cognitive radio networking," in *Proc. ICC*, Budapest, Hungary, Jun. 2013.
14. S. Hua, H. Liu, M. Wu, and S. Panwar, "Exploiting MIMO antennas in cooperative cognitive radio networks," in *Proc. INFOCOM*, pp. 2714–2722, Shanghai, China, Apr. 2011.
15. B. Cao, L. X. Cai, H. Liang, J. W. Mark, Q. Zhang, H. V. Poor, and W. Zhuang, "Cooperative cognitive radio networking using quadrature signaling," *Proc. INFOCOM*, pp. 3096–3100, Orlando, FL, Mar. 2012.
16. B. Cao, Q. Zhang, J. W. Mark, L. X. Cai, and H. V. Poor, "Toward efficient radio spectrum utilization: User cooperation in cognitive radio networking," *IEEE Network*, vol. 26, no. 4, pp. 46–52, Jul. 2012.
17. B. Cao, J. W. Mark, and Q. Zhang, "A polarization enabled cooperation framework for cognitive radio networking," in *Proc. GLOBECOM*, Anaheim, CA, Dec. 2012.
18. C. Oestges, B. Clerckx, M. Guillaud, and M. Debbah, "Dual-polarized wireless communications: From propagation models to system performance evaluation," *IEEE Trans. Wireless Commun.*, vol. 7, no. 10, pp. 4019–4031, Oct. 2008.
19. S. Kwon and G. L. Stüber, "Geometrical theory of channel depolarization," *IEEE Trans. Veh. Technol.*, vol. 60, no. 8, pp. 3542–3556, Oct. 2011.
20. B. Cao, Q. Zhang, L. Jin, and N. Zhang, "Oblique projection polarization filtering based interference suppressions for radar sensor networks", *EURASIP J. Wireless Commun. Netw.*, vol. 2010 (2010), Article ID 605103, 10 pages doi:10.1155/2010/605103.
21. Q. Zhang, B. Cao, J. Wang, and N. Zhang, "Polarization filtering technique based on oblique projections", *Sci. China Inf. Sci.*, vol. 53, no. 5, pp. 1056–1066, May 2010.
22. B. Cao, Q. Zhang, D. Liang, S. Wen, L. Jin, and Y. Zhang, "Blind adaptive polarization filtering based on oblique projection," in *Proc. ICC*, pp. 1–5, Cape Town, South Afica, May 2010.
23. V. Degli-Esposti, V.-M. Kolmonen, E. M. Vitucci, and P. Vainikainen, "Analysis and modeling on co- and cross-polarized urban radio propagation for dual-polarized MIMO wireless systems," *IEEE Trans. Antenn. Propag.*, vol. 59, no. 11, pp. 4247–4256, Nov. 2011.
24. F. Quitin, F. Bellens, A. Panahandeh, J.-M. Dricot, F. Dossin, F. Horlin, C. Oestges, and P. De Doncker, "A time-variant statistical channel model for tri-polarized antenna systems," *Proc. PIMRC*, pp. 64–69, Istanbul, Turkey, Sept. 2010.
25. R. Behrens and L. Scharf, "Signal processing applications of oblique projection operators," *IEEE Trans. Signal Processing*, vol. 42, no. 6, pp. 1413–1424, Jun. 1994.
26. H. Chang, P. Fard, S. Marcus, and M. Shayman, "Multitime scale Markov decision processes," *IEEE Trans. Autom. Control*, vol. 48, no. 6, pp. 976–987, Jun. 2003.

27. M. He, S. Murugesan, and J. Zhang, "Multiple timescale dispatch and scheduling for stochastic reliability in smart grids with wind generation integration," in *Proc. IEEE INFOCOM*, pp. 461–465, Shanghai, China, Apr. 2011.
28. S. Boyd and L. Vandenberghe, *Convex Optimization*. Cambridge, U.K.: Cambridge Univ. Press, 2003.
29. H. Chang, R. Givan, and E. Chong, "Parallel rollout for online solution of partially observable markov decision processes," *Discrete Event Dyna. Syst.: Theory Applicat.*, vol. 14, no. 3, pp. 309–341, 2004.
30. E. Chong, R. Givan, and H. Chang, "A framework for simulation-based network control via hindsight optimization," in *Proc. IEEE Conf. Decision Control*, pp. 1433–1438, Sydney, Austrlia, Dec. 2000.
31. D. Beyer, F. Cheng, S. Sethi, and M. Taksar, *Markovian Demand Inventory Models*. International Series in Operations Research and Management Science, vol. 108, New York: Springer, 2010.
32. H. Liang, B. J. Choi, W. Zhuang, and X. Shen, "Towards optimal energy store-carry-and-deliver for PHEVs via V2G system," in *Proc. INFOCOM*, pp. 1674–1682, Orlando, FL, Mar. 2012.
33. H. Shen, L. Cai, and X. Shen, "Performance analysis of TFRC over wireless link with truncated link-level ARQ," *IEEE Trans. Wireless Commun.*, vol. 5, no. 6, pp. 1479–1487, Jun. 2006.
34. L. E. Gurrieri, T. J. Willink, A. Petosa, and S. Noghanian, "Characterization of the angle, delay and polarization of multipath signals for indoor environments," *IEEE Trans. Antenn. Propag.*, vol. 56, no. 8, pp. 2710–2719, Aug. 2008.
35. H. S. Wang and N. Moayeri, "Finite-state Markov channel–A userful model for radio communication channels," *IEEE Trans. on Veh. Tech.*, vol. 44, no. 1, pp. 163–171, Feb. 1995.

Chapter 4
Optimal Communication Strategies in Cooperative Cognitive Radio Networking

Abstract This chapter is concerned with enhancement of spectrum utilization whereby a licensed primary user (PU) engages unlicensed secondary users (SUs) to relay its transmission in an energy-aware cognitive radio network to expedite information transfer. The cooperation can be pure relaying or provide diversity transmissions using an amplify-and-forward or decode-and-forward mode. In a cooperative cognitive radio network (CCRN), the individual cooperating partner attempts to maximize its own utility. The energy-aware partner selection and parameter optimization process, led by the PU, is formulated as two Stackelberg games, namely a sum-constrained power allocation game for two-phase and a power control game for three-phase cooperation, respectively. Unique Nash Equilibrium is proved and achieved in analytical format for each game. The optimal communication strategy is chosen which achieves the maximum PU utility among different optimal communication strategies. Moreover, an implementation scheme is presented to perform the partner selection and parameter optimization based on the analytical results. Theoretical analysis and performance evaluation show that the proposed CCRN model is a promising framework under which the PU's utility is maximized, while the relaying SUs can attain acceptable utilities.

Cognitive radios have attracted extensive interests in addressing radio spectrum efficiency and improving utilization in the context of current static spectrum allocation policies [1–5]. In cognitive radio networks, unlicensed secondary users (SUs) are allowed to access the licensed spectrum as long as SUs' transmissions cause negligible interference to primary users (PUs). Therefore, an SU can (i) opportunistically access the temporarily unused spectrum via spectrum sensing, (ii) concurrently access the licensed spectrum without disrupting PUs' transmissions, or (iii) cooperatively obtain transmission opportunities through collaborating with PUs [1, 6]. In [1], these access models for cognitive radio networks are termed interweave, underlay, and overlay, respectively.

The focus of this chapter is on SUs leveraging the overlay access model to enable cooperation with PUs, i.e., forming a cooperative cognitive radio network (CCRN) [6–14]. In CCRN, PUs employ SUs for helping relay PUs' traffic to

© The Author(s) 2016
B. Cao et al., *Cooperative Cognitive Radio Networking*, SpringerBriefs in Electrical and Computer Engineering, DOI 10.1007/978-3-319-32881-2_4

improve primary transmissions, and in return lend a portion of spectrum access time to the relaying SUs for secondary transmissions [7], which creates a win-win situation.

There have been extensive studies on CCRN [8–14]. In [8], PUs and SUs establish TDMA based three-phase cooperation, i.e., PUs broadcast in the first phase, SUs relay in the second phase, and SUs transmit in the third phase. In [9], the authors propose an FDMA based two-phase cooperation, in which a PU divides its spectrum into two orthogonal subbands, and broadcasts on the first subband in the first phase. SUs relay on the same subband in the second phase, and continuously transmit in both two phases on the second subband. In [10], an MIMO enabled two-phase cooperation is presented. By leveraging the degrees of freedom provided by MIMO, SUs concurrently relay and transmit in the second phase. In [11], the authors investigate the use of quadrature modulation for CCRN to attain two-phase cooperation, i.e., SUs simultaneously relay for PUs using the quadrature channel while transmit their own using the in-phase channel.

In regard to evaluating performance of CCRN, utility optimization for some specific communication strategy is considered, however, these results may not be applicable to more generalized cases. In addition, impacts of energy and power consumptions on communication strategies attract little attention in CCRN. Motivated by these problems, we aim to find the optimal communication strategies and the associated optimal parameters among multiple communication strategies, by taking energy and power consumptions into consideration.

In our framework, PUs can improve throughput by either increasing transmit power in direct transmission, or exploiting SUs to build CCRN, in which SUs can provide multihop or cooperative relay for PUs,[1] as shown in Fig. 4.1. In both multihop and cooperative relay scenarios, PUs and SUs can collaborate in a two-phase or three-phase manner, as shown in Fig. 4.2. SUs can relay for PUs using amplify-and-forward (AF) or decode-and-forward (DF). In this context, there are nine different communications strategies: (i) direct transmission, (ii) two-phase AF multihop, (iii) two-phase DF multihop, (iv) three-phase AF multihop, (v) three-phase DF multihop, (vi) two-phase AF cooperative, (vii) two-phase DF cooperative, (viii) three-phase AF cooperative, (ix) three-phase DF cooperative, respectively. For each communication strategy, optimal relay/transmit powers need to be found.

With the emphasis on energy and power impacts, we target at maximizing PUs' utilities while achieving acceptable utilities for SUs through finding optimal communication strategies and the associated optimal parameters. In this chapter, utility functions which incorporate quantitative impacts of energy and power consumptions on utility cost, are performance metrics reflecting spectrum-energy efficiency. This optimization problem is solved by maximizing PUs utilities for each

---

[1]Actually these two terms both indicate cooperation, however, we use multihop relay to represent the pure relaying scenario, and cooperative relay to capture the nature that direct link between the transmitter and receiver is available. We hope this usage does not introduce confusion.

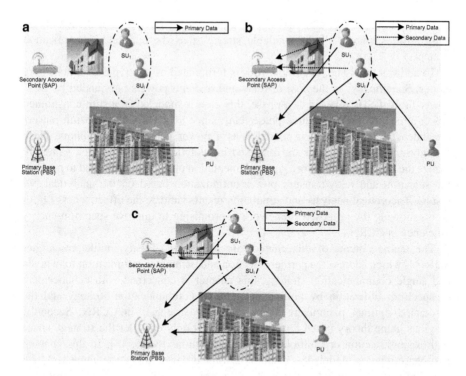

Fig. 4.1 Different transmission strategies. (**a**) Direct transmission. (**b**) Multihop relay. (**c**) Cooperative relay

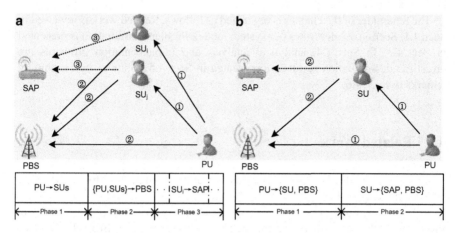

Fig. 4.2 SUs' relaying strategies for a PU. (**a**) Three-phase relay. (**b**) Two-phase relay

communication strategy and then choosing the strategy that provides the maximum PUs' utilities among all the available strategies as the optimal communication strategy.

To address this issue, our framework is formulated as two types of Stackelberg games. Specifically, as the sum of relay and transmit powers is bounded by SUs' power limit if SUs relay in two-phase, this case is modeled as a sum-constrained power allocation game. In three-phase relay, since SUs' relay and transmit powers are independent, this case is modeled as a power control game. Unique Nash Equilibrium (NE) is proved and achieved in analytical format for each game. To realize the proposed framework, an implementation protocol is presented to perform SU selection and relay/transmit power optimization based on the analytical NE results. Theoretical analysis and simulation results validate the effectiveness of our work, showing the proposed framework is promising to improve spectrum-energy efficiency in CCRN.

The main contents of this chapter are as follows. Firstly, unlike researches in [8–14] which address the partner selection and parameter optimization for a given and single communication strategy, this chapter is concerned with enhancement of spectrum utilization by finding the optimal communication strategy and the associated optimal parameters among multiple strategies in CCRN. Secondly, previous game theory based results about CCRN only consider the strategy space with a single action or multiple independent actions [8, 10, 13]. In this chapter, two-phase relay is modeled as a sum-constrained power allocation game with two bounded powers coupled together. Thirdly, previous work does not consider the impact of energy and power on optimal strategy selection. In this chapter, we emphasize on this effect by incorporating energy level and power consumptions into utility function design.

The remainder of the chapter is organized as follows. Related work is reviewed in Sect. 4.1. Section 4.2 describes our system model. Problem formulation is presented in Sect. 4.3. In Sect. 4.4, analytical analysis and implementation protocols are investigated. Simulation results are presented in Sect. 4.5, followed by concluding remarks in Sect. 4.6.

4.1 Related Work

Game theory based performance analysis and utility optimization for CCRN have been extensively studied.

In [8], the authors propose a payment mechanism in which SUs pay charges to PUs in CCRN. The model is formulated as a non-cooperative Stackelberg game to maximize PUs' and SUs' utilities in three-phase DF cooperation, and a unique NE is proved in SUs' payment strategies. This work assumes that both PUs and SUs transmit and relay using constant powers, and the revenue is a constant. In addition, no throughput constraints and energy issues are considered. In [10], an MIMO-CCRN framework is proposed to enable two-phase DF cooperation between PUs and SUs. The authors model the optimization problem as a Stackelberg game, and derive the optimal phase durations and relay selection based on NE in optimal relay powers of SUs, while SUs' transmit powers are constants. In [12], to address how such cooperation can be exploited in OFDMA based CCRN and the selfishness feature in resource allocation, the authors formulate an optimization framework based on Nash Bargaining Solutions to fairly and efficiently allocate resource between PUs and SUs. All the above work is based on the assumption that cooperation is always beneficial to PUs, which maybe invalid under some conditions. In [13], by observing the effectiveness of cooperation, the authors consider the problems of when to cooperate and how to cooperate in three-phase DF CCRN, and the optimization problem is formulated as a Stackelberg game.

Previous work only takes account of utility optimization for a given communication strategy, and does not consider the impacts of energy and power on optimal communication strategy. Based on quantifying impacts of energy and power on utility functions, we aim to find the optimal communication strategy among multiple communication strategies.

4.2 System Model

In CCRN, there are one primary network consisting of PUs and a primary base station (PBS), and one secondary network consisting of SUs and a secondary access point (SAP). We consider PUs and SUs are imposed by energy and power constraints, and work in half-duplex mode. A PU has three possible transmission strategies: *direct transmission*, *multihop relay*, and *cooperative relay*, respectively, as shown in Fig. 4.1. In either multihop or cooperative relay, SUs can forward PU's data and transmit their own information in either a *two-phase* or *three-phase* manner. Take the three-phase cooperative transmission strategy for an instance, as shown in Fig. 4.1c and Fig. 4.2a, the PU repeatedly broadcasts its signals in the first and second phase, SUs forward PU's data to the PBS in the second phase, and access PU's spectrum for their own transmissions in a TDMA mode during the third phase, respectively. In addition, for either two-phase or three-phase relay, SUs can use AF or DF to forward PU's data.

Define $l \in \{M, C\}$ as the set of transmission strategies with cooperation, i.e., M stands for multihop, $m \in \{2, 3\}$ as the set of relay strategies, and $n \in \{A, D\}$ as the set of forwarding strategies, respectively. In this context, the communication strategy set $\{l, m, n\}$ is the communication strategy in terms of transmission-relay-forwarding schemes. Each SU is required to use the same communication strategy with other SUs when they are involved in relaying PU's data so that the destination of PU can effectively and simply combine the relayed paths from SUs, while the relaying SUs can user different relay/transmit powers. All the relaying SUs form a set $\mathscr{S}$ with size N, i.e., $|\mathscr{S}| = N$.

Without loss of generality, the bandwidth and length of time slot allocated to PUs are both normalized to be one Hertz and one second, respectively. Hence the term spectrum efficiency and throughput are used interchangeably. The duration of each phase in both relay strategies are equally divided, i.e., each phase duration is $\frac{1}{2}$ for two-phase, and $\frac{1}{3}$ for three-phase.

Given the set $\{l, m, n\}$, we denote P_{Plmn} as PU's transmit power, P_{S_iPlmn} as the power SU$_i$ uses for relaying, and P_{S_ilmn} as SU$_i$'s power for its own transmission, where $S_i \in \mathscr{S}$. To capture the power constrained feature, we have $P_{Plmn} \leq P_{PM}$, $P_{S_il3n} \leq P_{S_iM}$ and $P_{S_iPl3n} \leq P_{S_iM}$ for three-phase, and $P_{S_il2n} + P_{S_iPl2n} \leq P_{S_iM}$ for two-phase, where P_{PM} and P_{S_iM} are power limits of the PU and SU$_i$, respectively. The PU uses a power level P_{PD} if it communicates in direct transmission without cooperation.

The channels are modeled as independent proper complex Gaussian random variables invariant within one slot, e.g., Rayleigh block-fading. We use Γ_{PP}, Γ_{PS_i}, Γ_{S_iP}, and Γ_{SS_i} to represent the channel gain to noise ratio, e.g., $\Gamma_{PP} = \frac{|h_{PP}|^2}{\sigma^2}$ with h_{PP} being the instantaneous channel fading coefficient and σ^2 being the power of additive noise.

Denote ω_{S_ilmn} as the fraction of time that SU$_i$ can gain for its own transmission during the total secondary transmissions. We assume ω_{S_ilmn} is proportional to SU$_i$'s power consumed for relaying PU's signal, i.e.,

$$\omega_{S_ilmn} = \frac{P_{S_iPlmn}}{\sum_{j=1}^{N} P_{S_jPlmn}}, \quad S_j \in \mathscr{S}. \tag{4.1}$$

Therefore, the time duration SU$_i$ gains for its own data transmission is $\frac{\omega_{S_ilmn}}{m}$ in m-phase relaying.

4.3 Problem Formulation

In this section, we define utility functions and model the problem as two types of Stackelberg games.

4.3.1 Utility Functions

PUs aim to maximize their utilities over different communication strategies. Given a PU's energy level E_{P0}, the spectrum-energy efficiency based utility function is defined as

$$U_P = \begin{cases} U_{PD} = R_{PD} - J(E_{P0})E_{PD} \\ U_{Plmn} = R_{Plmn} - J(E_{P0})E_{Plmn} \end{cases} \tag{4.2}$$

where U_{PD} and U_{Plmn} are PU's utilities without and with cooperation; R_{PD} and R_{Plmn} are PU's achievable rates without and with cooperation; E_{PD} and E_{Plmn} are PU's energy consumptions without and with cooperation, respectively. A nonnegative

$J(E_{P0})$ is the per energy consumption penalty which is a function of PU's initial energy level E_{P0}. It is reasonable to define $J(E_{P0})$ as a decreasing function of E_{P0}, since given a fixed energy consumption E_{Plmn}, the more the initial energy is, the less E_{Plmn} impacts.

Since PUs have higher priority over SUs, PUs are leaders when cooperating with SUs, i.e., PUs decide whether to cooperate and how to cooperate. Specifically, a PU determines P_{PD} if not cooperating, and determines the values of set $\{l, m, n, P_{Plmn}\}$ if cooperating with SUs. In this context, PUs target at maximizing constrained utilities as

$$\textbf{(P1)} \quad \max_{\{l,m,n,P_{Plmn}\}/P_{PD}} \quad U_P \tag{4.3}$$

$$\text{Subject to :} \quad R_P \geq R_{PT}$$

$$0 < P_{Plmn}, P_{PD} \leq P_{PM}$$

where R_P and R_{PT} are PU's actual and expected throughput.

Problem **(P1)** can be solved by comparing maximum utility in direct transmission and maximum utility in cooperation,

$$\max \left\{ U_{PD}(P_{PD}^*), U_{Pl^*m^*n^*}(P_{Plmn}^*) \right\} \tag{4.4}$$

where $U_{PD}(P_{PD}^*)$ is PU's maximum utility under optimal transmit power P_{PD}^* in direct transmission, and the set

$$\{l^*, m^*, n^*, P_{Plmn}^*\} = \arg \max_{\substack{l \in \{M,C\}, m \in \{2,3\}, \\ n \in \{A,D\}, P_{Plmn} \in (0,P_{PM}]}} U_{Plmn} \tag{4.5}$$

is optimal communication and power strategy in cooperation.

SU$_i$ aims to maximize the following utility function

$$\textbf{(P2)} \quad \max_{P_{S_ilmn}, P_{S_iPlmn}} \quad U_{S_ilmn} = \epsilon_i(P_{S_ilmn})R_{S_ilmn} \tag{4.6}$$

$$- \vartheta_i(P_{S_iPlmn})E_{S_iPlmn}$$

$$\text{Subject to :} \quad R_{S_ilmn} \geq R_{S_iT}$$

$$0 < P_{S_il3n}, P_{S_iPl3n} \leq P_{S_iM}$$

$$0 < P_{S_il2n} + P_{S_iPl2n} \leq P_{S_iM}$$

where $\epsilon_i(P_{S_ilmn})$ is the equivalent revenue per unit throughput that contributes to U_{S_ilmn}. $\epsilon_i(P_{S_ilmn})$ is continuous, twice differentiable, positive and nonincreasing concave with respect to (w.r.t.) P_{S_ilmn}. $\vartheta_i(P_{S_iPlmn})$ is the equivalent cost per unit relaying energy consumption to the overall utility, and is continuous, twice differentiable, positive and nondecreasing convex w.r.t. P_{S_iPlmn}. The rationale behind the assumptions of revenue and cost functions is that, given the per unit achievable throughput, SU$_i$ obtains more revenue if it consumes less energy; the more energy SU$_i$ uses to relay, the greater cost the cooperation introduces to SU$_i$. Definitions about concavities of $\epsilon_i(P_{S_ilmn})$ and $\vartheta_i(P_{S_iPlmn})$ are widely applied in modeling revenue and cost functions, such as the Sigmoidal function in [8]. The throughput constraint shows SU_i's throughput R_{S_ilmn} should not be less than a certain threshold R_{S_iT}.

The PBS combines the multipath primary transmissions using maximal ratio combining, and PUs' and SUs' throughput and energy consumptions for each strategy are summarized in Fig. 4.3, where $f(x,y) = \frac{xy}{x+y+1}$ and $C(x)$ is the achievable Shannon capacity under the signal to noise ratio x. For the sake of computation simplicity, we define $C(x) = \ln(1+x)$.

4.3.2 Game Theoretic Analysis

Given $\{l, m, n, P_{Plmn}\}$ by the PU, U_{S_ilmn} is a function w.r.t. P_{S_ilmn} and P_{S_iPlmn}. Specifically, for $m = 2$, P_{S_il2n} and P_{S_iPl2n} are coupled together due to power sum constraint. Thus a non-cooperative sum-constrained power allocation game (AG) [15] w.r.t. the pairwise power allocation vector $(P_{S_il2n}, P_{S_iPl2n})$ is formulated for two-phase relay. In three-phase relay, P_{S_il3n} and P_{S_iPl3n} are independent, hence a non-cooperative power control game (CG) is modeled for $m = 3$.

Direct			$R_{PD} - C(P_{PD}\Gamma_{PP}), E_{PD} - P_{PD}$
Multihop	2-P	AF:	$R_{PM2A} = \frac{1}{2}C\left(\sum_{i=1}^{N} f(P_{PM2A}\Gamma_{PS_i}, P_{S_iPM2A}\Gamma_{S_iP})\right), E_{PM2A} = \frac{1}{2}P_{PM2A}$
			$R_{S_iM2A} = \frac{\omega_{S_iM2A}}{2}C(P_{S_iM2A}\Gamma_{SS_i}), E_{S_iM2A} = \frac{1}{2}(P_{S_iPM2A} + \omega_{S_iM2A}P_{S_iM2A})$
		DF:	$R_{PM2D} = \frac{1}{2}\min\left\{C\left(\min_{k\in\mathscr{S}}P_{PM2D}\Gamma_{PS_k}\right), C\left(\sum_{i=1}^{N}P_{S_iPM2D}\Gamma_{S_iP}\right)\right\}, E_{PM2D} = \frac{1}{2}P_{PM2D}$
			$R_{S_iM2D} = \frac{\omega_{S_iM2D}}{2}C(P_{S_iM2D}\Gamma_{SS_i}), E_{S_iM2D} = \frac{1}{2}(P_{S_iPM2D} + \omega_{S_iM2D}P_{S_iM2D})$
	3-P	AF:	$R_{PM3A} = \frac{1}{3}C\left(\sum_{i=1}^{N} f(P_{PM3A}\Gamma_{PS_i}, P_{S_iPM3A}\Gamma_{S_iP})\right), E_{PM3A} = \frac{1}{3}P_{PM3A}$
			$R_{S_iM3A} = \frac{\omega_{S_iM3A}}{3}C(P_{S_iM3A}\Gamma_{SS_i}), E_{S_iM3A} = \frac{1}{3}(P_{S_iPM3A} + \omega_{S_iM3A}P_{S_iM3A})$
		DF:	$R_{PM3D} = \frac{1}{3}\min\left\{C\left(\min_{k\in\mathscr{S}}P_{PM3D}\Gamma_{PS_k}\right), C\left(\sum_{i=1}^{N}P_{S_iPM3D}\Gamma_{S_iP}\right)\right\}, E_{PM3D} = \frac{1}{3}P_{PM3D}$
			$R_{S_iM3D} = \frac{\omega_{S_iM3D}}{3}C(P_{S_iM3D}\Gamma_{SS_i}), E_{S_iM3D} = \frac{1}{3}(P_{S_iPM3D} + \omega_{S_iM3D}P_{S_iM3D})$
Cooperative	2-P	AF:	$R_{PC2A} = \frac{1}{2}C\left(P_{PC2A}\Gamma_{PP} + \sum_{i=1}^{N} f(P_{PC2A}\Gamma_{PS_i}, P_{S_iPC2A}\Gamma_{S_iP})\right), E_{PC2A} = P_{PC2A}$
			$R_{S_iC2A} = \frac{\omega_{S_iC2A}}{2}C(P_{S_iC2A}\Gamma_{SS_i}), E_{S_iC2A} = \frac{1}{2}(P_{S_iPC2A} + \omega_{S_iC2A}P_{S_iC2A})$
		DF:	$R_{PC2D} = \frac{1}{2}\min\left\{C\left(\min_{k\in\mathscr{S}}P_{PC2D}\Gamma_{PS_k}\right), C\left(P_{PC2D}\Gamma_{PP} - \sum_{i=1}^{N}P_{S_iPC2D}\Gamma_{S_iP}\right)\right\}, E_{PC2D} = P_{PC2D}$
			$R_{S_iC2D} = \frac{\omega_{S_iC2D}}{2}C(P_{S_iC2D}\Gamma_{SS_i}), E_{S_iC2D} = \frac{1}{2}(P_{S_iPC2D} + \omega_{S_iC2D}P_{S_iC2D})$
	3-P	AF:	$R_{PC3A} = \frac{1}{3}C\left(P_{PC3A}\Gamma_{PP} + \sum_{i=1}^{N} f(P_{PC3A}\Gamma_{PS_i}, P_{S_iPC3A}\Gamma_{S_iP})\right), E_{PC3A} = \frac{2}{3}P_{PC3A}$
			$R_{S_iC3A} = \frac{\omega_{S_iC3A}}{3}C(P_{S_iC3A}\Gamma_{SS_i}), E_{S_iC3A} = \frac{1}{3}(P_{S_iPC3A} + \omega_{S_iC3A}P_{S_iC3A})$
		DF:	$R_{PC3D} = \frac{1}{3}\min\left\{C\left(\min_{k\in\mathscr{S}}P_{PC3D}\Gamma_{PS_k}\right), C\left(P_{PC3D}\Gamma_{PP} - \sum_{i=1}^{N}P_{S_iPC3D}\Gamma_{S_iP}\right)\right\}, E_{PM3D} = \frac{2}{3}P_{PC3D}$
			$R_{S_iC3D} = \frac{\omega_{S_iC3D}}{3}C(P_{S_iC3D}\Gamma_{SS_i}), E_{S_iC3D} = \frac{1}{3}(P_{S_iPC3D} + \omega_{S_iC3D}P_{S_iC3D})$

Fig. 4.3 System throughput and energy consumptions of the PU and SU for different communication strategies.

Let $AG = [\mathscr{S}, \mathscr{P}_S^{AG}, \mathscr{U}_S]$ represent the non-cooperative power allocation game for two-phase relay, where $\mathscr{S} = \{S_1, \ldots, S_i, \ldots S_N\}$ is the set of cooperating SUs; $\mathscr{P}_S^{AG} = \{\mathscr{P}_{Sl2n}^{AG}, \mathscr{P}_{SPl2n}^{AG}\}$ with $\mathscr{P}_{Sl2n}^{AG} = \times_{S_i \in \mathscr{S}} P_{S_i l2n}^{AG}$ and $\mathscr{P}_{SPl2n}^{AG} = \times_{S_i \in \mathscr{S}} P_{S_i Pl2n}^{AG}$ are transmit and relay power sets; $\mathscr{U}_S = \times_{S_i \in \mathscr{S}} U_{S_i l2n}$ is the set of utility functions. In particular, SU$_i$'s utility function $U_{S_i l2n} : \mathscr{P}_{S_i}^{AG} \to \mathscr{R}$ generally depends on the power strategies $\mathscr{P}_S^{AG} = (\mathscr{P}_{S_i}^{AG}, \mathscr{P}_{-S_i}^{AG})$ of all SUs, where $\mathscr{P}_{S_i}^{AG}$ represents a feasible power allocation action set of SU$_i$ for $S_i \in \mathscr{S}$, and $\mathscr{P}_{-S_i}^{AG}$ is a pairwise vector of the power allocation actions of other SUs except SU$_i$ in the non-cooperative AG. Since AG happens in two-phase relay, we omit the superscript AG in each term when $m = 2$ for simplicity.

The non-cooperative power control game is represented as $CG = [\mathscr{S}, \mathscr{P}_S^{CG}, \mathscr{U}_S]$ for three-phase relay, and the strategy space $\mathscr{P}_{Sl3n} = \times_{S_i \in \mathscr{S}} P_{S_i l3n}$ and $\mathscr{P}_{SPl3n} = \times_{S_i \in \mathscr{S}} P_{S_i Pl3n}$ are relay and transmit power sets. In particular, SU$_i$'s utility function $U_{S_i l3n} : \mathscr{P}_{S_i} \to \mathscr{R}$ generally depends on the power strategies $\mathscr{P}_S = (\mathscr{P}_{S_i}, \mathscr{P}_{-S_i})$ of all SUs.

In a non-cooperative game, a NE is a fixed point of the game in which no user can increase its utility through its own actions. Before we investigate NE solutions for the proposed AG and CG problems, the following definition of NE for the non-cooperative AG is given [16].

Definition 3. A pairwise power allocation vector $\mathscr{P}_S^* = \{\mathscr{P}_{Sl2n}^*, \mathscr{P}_{SPl2n}^*\}$ is a NE of the non-cooperative $AG = [\mathscr{S}, \mathscr{P}_S^{AG}, \mathscr{U}_S]$, if $\forall\ S_i \in \mathscr{S}$ and $P_{S_i} \in \mathscr{P}_{S_i}$; we always have

$$U_{S_i l2n}(P_{S_i}^*, \mathscr{P}_{-S_i}^*) \geq U_{S_i l2n}(P_{S_i}, \mathscr{P}_{-S_i}^*). \tag{4.7}$$

Definition of NE for CG can be obtained similarly.

4.4 Stackelberg Game Framework Analysis

In this section, we decompose and analyze the proposed optimization problem by using typical two-stage Stackelberg game. We address the unique NE for the non-cooperative AG and CG and maximize PU's utility based on the obtained NE.

4.4.1 NE Analysis for AG

In (**P2**), the first constraint can be satisfied by solving the throughput inequality which yields the cooperation region as discussed in [11]. The cooperation region represents a threshold of Γ_{SS_i}. Given an expected throughput, the inequality holds when Γ_{SS_i} is above the threshold, then (**P2**) in two-phase relay can be equivalently rewritten as

$$(\textbf{P3}) \quad \max_{P_{S_i l2n}, P_{S_i Pl2n}} \quad U_{S_i l2n} = \frac{\epsilon_i P_{S_i Pl2n} C(P_{S_i l2n} \Gamma_{SS_i})}{2\sum_{j=1}^{N} P_{S_j Pl2n}} \tag{4.8}$$

$$- \frac{\vartheta_i P_{S_i Pl2n}}{2}$$

$$\text{Subject to}: 0 < P_{S_i l2n} + P_{S_i Pl2n} \leq P_{S_i}.$$

Theorem 2. *A NE in relay power strategy $\mathscr{P}_{SPl2n}$ for the non-cooperative AG $= [\mathscr{S}, \mathscr{P}_S^{AG}, \mathscr{U}_S]$ exists and is unique.*

Proof 6. The following two propositions are used to prove the existence and uniqueness of the NE in Theorem 2.

Proposition 1. *A NE exists in the non-cooperative AG, if for $S_i \in \mathscr{S}$, the following properties hold true: (i) the strategy on relay power $\mathscr{P}_{SPl2n} = \times_{S_i \in \mathscr{S}} P_{S_i Pl2n}$ is a nonempty, convex and compact subset of some Euclidean space $\mathfrak{R}^N$; (ii) SU_i's utility function $U_{S_i l2n}(P_{S_i Pl2n}, \mathscr{P}_{-S_i Pl2n})$ is continuous in $\mathscr{P}_{SPl2n} = (P_{S_i Pl2n}, \mathscr{P}_{-S_i Pl2n})$ and concave in $P_{S_i Pl2n}$, where $\mathscr{P}_{-S_i} = (P_{S_1}, \ldots, P_{S_{i-1}}, P_{S_{i+1}}, \ldots, P_{S_N})$ [8].*

Proposition 2. *The NE in $\mathscr{P}_{SPl2n}$ for the non-cooperative AG is unique if the weighted sum of utility functions with nonnegative weight vector $\gamma = (\gamma_1, \ldots \gamma_i, \ldots \gamma_N)$, i.e.,*

$$\mu(\mathscr{P}_{SPl2n}, \boldsymbol{\gamma}) = \sum_{i=1}^{N} \gamma_i U_{S_i l2n}(P_{S_i Pl2n}, \mathscr{P}_{-S_i Pl2n}) \tag{4.9}$$

for all $S_i \in \mathscr{S}$ and $\gamma_i \geq 0$, is diagonally strictly concave [16].

According to the physical interpretation of SU_i's relay power $P_{S_i Pl2n} \in (0, P_{S_i M})$, $\mathscr{P}_{SPl2n}$ is a nonempty, convex and compact subset of a positive Euclidean space $\mathfrak{R}_+^N$. It is easy to verify the utility function $U_{S_i l2n}$ is continuous w.r.t. both $P_{S_i Pl2n}$ and $\mathscr{P}_{-S_i Pl2n} = (P_{S_1 Pl2n}, \ldots, P_{S_{i-1} Pl2n}, P_{S_{i+1} Pl2n}, \ldots, P_{S_N Pl2n})$. Therefore, $U_{S_i l2n}$ is continuous in $\mathscr{P}_{SPl2n} = (P_{S_i Pl2n}, \mathscr{P}_{-S_i Pl2n})$.

We now calculate the second-order derivative w.r.t. $P_{S_i Pl2n}$ to check the concavity of $U_{S_i l2n}$, which is expressed as

$$\frac{\partial^2 U_{S_i l2n}}{\partial^2 P_{S_i Pl2n}} = - \frac{\epsilon_i C(P_{S_i l2n} \Gamma_{SS_i}) \sum_{j \neq i} P_{S_j Pl2n}}{\left(\sum_{j=1}^{N} P_{S_j Pl2n}\right)^3} \tag{4.10}$$

$$- \frac{\vartheta_i'' P_{S_i Pl2n} + 2\vartheta_i'}{2}.$$

Since $\epsilon_i > 0$, $\vartheta_i' \geq 0$ and $\vartheta_i'' \geq 0$ according to our definitions, $\frac{\partial^2 U_{S_i l2n}}{\partial^2 P_{S_i Pl2n}}$ in equation (4.10) is always negative, thus SU_i's utility function $U_{S_i l2n}$ is strictly concave in $P_{S_i Pl2n}$.

This ensures the existence of a NE in relay power strategy $\mathscr{P}_{SPl2n}$ for the non-cooperative AG $= [\mathscr{S}, \mathscr{P}_S^{AG}, \mathscr{U}_S]$.

To verify the uniqueness of NE, we examine the properties of the weighted sum utilities with nonnegative weights. One equivalent way to prove $\mu(\mathscr{P}_{SPl2n}, \boldsymbol{\gamma})$ to be diagonally strictly concave is by investigating the properties about the Jacobian matrix of its pseudo-gradient. Before we prove the uniqueness of NE, the pseudo-gradient of the weighted sum of utility functions associated with its Jacobian matrix, and the following lemma which has been proved in [17] are presented.

The pseudo-gradient of equation (4.9) is given by

$$\mathscr{G}(\mathscr{P}_{SPl2n}, \boldsymbol{\gamma}) = [\gamma_1 \nabla U_{S_1 l2n}, \ldots, \gamma_N \nabla U_{S_N l2n}]^T \tag{4.11}$$

where $\nabla U_{S_i l2n} = \frac{\partial U_{S_i l2n}}{\partial P_{S_i Pl2n}}$ for $S_i \in \mathscr{S}$, and the superscript T denotes the vector transpose.

The Jacobian matrix $\mathbf{J}(\mathscr{P}_{SPl2n}, \boldsymbol{\gamma})$ w.r.t. $\mathscr{P}_{SPl2n}$ of the pseudo-gradient $\mathscr{G}(\mathscr{P}_{SPl2n}, \boldsymbol{\gamma})$ is the matrix with its entry J_{ij} ($1 \leq i, j \leq N$) obtained as

$$J_{ij} = \begin{cases} \gamma_i \dfrac{\partial(\nabla U_{S_i l2n})}{\partial P_{S_i Pl2n}} = \gamma_i \dfrac{\partial^2 U_{S_i l2n}}{\partial^2 P_{S_i Pl2n}}, & \text{for } i = j; \\[4mm] \gamma_i \dfrac{\partial(\nabla U_{S_i l2n})}{\partial P_{S_j Pl2n}} = \gamma_i \dfrac{\partial^2 U_{S_i l2n}}{\partial P_{S_i Pl2n} \partial P_{S_j Pl2n}}, & \text{for } i \neq j. \end{cases}$$

Lemma 5. For $U_{S_i l2n}(P_{S_i}, \mathscr{P}_{-S_i})$ which is strictly concave w.r.t. P_{S_i} and convex w.r.t. $\mathscr{P}_{-S_i}$, if there is some nonnegative vector $\boldsymbol{\gamma}$ such that $\mu(\mathscr{P}_{SPl2n}, \boldsymbol{\gamma})$ is concave w.r.t. $\mathscr{P}_{SPl2n}$, then the matrix $[\mathbf{J}(\mathscr{P}_{SPl2n}, \boldsymbol{\gamma}) + \mathbf{J}^T(\mathscr{P}_{SPl2n}, \boldsymbol{\gamma})]$ is negative definite, where $\mathbf{J}(\mathscr{P}_{SPl2n}, \boldsymbol{\gamma})$ is the Jacobian matrix of $\mathscr{G}(\mathscr{P}_{SPl2n}, \boldsymbol{\gamma})$.

We have proved that $U_{S_i l2n}(P_{S_i}, \mathscr{P}_{-S_i})$ is strictly concave w.r.t. $P_{S_i Pl2n}$ according to (4.10). We now take the second-order derivative w.r.t. $P_{S_j Pl2n}$ for $S_j \in \mathscr{S}$ and $j \neq i$ to prove the convexity, which is obtained as

$$\frac{\partial^2 U_{S_i l2n}}{\partial^2 P_{S_j Pl2n}} = \frac{\epsilon_i P_{S_i Pl2n} C(P_{S_i l2n} \Gamma_{SS_i})}{(\sum_{j=1}^N P_{S_j Pl2n})^3} > 0. \tag{4.12}$$

The second-order derivative of $U_{S_i l2n}$ w.r.t. $P_{S_j Pl2n}$ for $S_j \in \mathscr{S}$ and $j \neq i$ is always larger than 0. Therefore, $U_{S_i l2n}(P_{S_i}, \mathscr{P}_{-S_i})$ is convex w.r.t. $\mathscr{P}_{-S_i}$.

We take the second-order derivative of $\mu(\mathscr{P}_{SPl2n}, \boldsymbol{\gamma})$ w.r.t. $\mathscr{P}_{SPl2n}$, which yields

$$\frac{\partial^2 \mu(\mathscr{P}_{SPl2n}, \boldsymbol{\gamma})}{\partial^2 P_{S_i Pl2n}} = -\sum_{i=1}^N \gamma_i \left(\frac{\vartheta_i'' P_{S_i Pl2n} + 2\vartheta_i'}{2} \right)$$

$$- \sum_{i=1}^N \gamma_i \frac{\epsilon_i C(P_{S_i l2n} \Gamma_{SS_i}) \sum_{j \neq i}^N P_{S_j Pl2n}}{(\sum_{j=1}^N P_{S_j Pl2n})^3} < 0. \tag{4.13}$$

This guarantees $\mu(\mathscr{P}_{SPl2n}, \boldsymbol{\gamma})$ is concave w.r.t. $\mathscr{P}_{SPl2n}$.

Based on Lemma 5, $[\mathbf{J}(\mathscr{P}_{SP l2n}, \boldsymbol{\gamma}) + \mathbf{J}^T(\mathscr{P}_{SP l2n}, \boldsymbol{\gamma})]$ is negative definite, showing that the weighted sum of utility functions with nonnegative weight vector is diagonally strictly concave. According to Proposition 2, the NE in $\mathscr{P}_{SP l2n}$ for the non-cooperative AG is unique.

Therefore, there is a unique NE exists in $\mathscr{P}_{SP l2n}$ for the non-cooperative $AG = [\mathscr{S}, \mathscr{P}_S^{AG}, \mathscr{U}_S]$.

We analyze the NE in SUs' transmit power strategy for their own data transmissions in the non-cooperative power allocation game by proving the following theorem.

Theorem 3. *A unique NE exists in the transmit power strategy $\mathscr{P}_{Sl2n}$ for the non-cooperative $AG = [\mathscr{S}, \mathscr{P}_S^{AG}, \mathscr{U}_S]$.*

Proof 7. Note that SU_i's strategy space $\mathscr{P}_{S_i l2n}$ is a nonempty, convex and compact subset of some Euclidean space $\mathfrak{R}^N$. In addition, the concavity can be proved by calculating the second-order derivative of SU_i's utility function $U_{S_i l2n}$ w.r.t. $P_{S_i l2n}$, i.e.,

$$\frac{\partial^2 U_{S_i l2n}}{\partial^2 P_{S_i l2n}} = \frac{\epsilon_i'' P_{S_i Pl2n} C(P_{S_i l2n} \Gamma_{SS_i})}{2 \sum_{j=1}^N P_{S_j Pl2n}}$$

$$+ \frac{\epsilon_i' P_{S_i Pl2n} \Gamma_{SS_i}}{(1 + P_{S_i l2n}) \sum_{j=1}^N P_{S_j Pl2n}} \qquad (4.14)$$

$$- \frac{\epsilon_i P_{S_i Pl2n} \Gamma_{SS_i}^2}{2(1 + P_{S_i l2n})^2 \sum_{j=1}^N P_{S_j Pl2n}}.$$

Since we have $\epsilon_i > 0$, $\epsilon_i' \le 0$ and $\epsilon_i'' \le 0$ by definition, $\frac{\partial^2 U_{S_i l2n}}{\partial^2 P_{S_i l2n}}$ is always negative, thus $U_{S_i l2n}$ is strictly concave in $P_{S_i l2n}$. This ensures the existence of NE in $\mathscr{P}_{S l2n}$.

In order to prove uniqueness, we define the weighted sum of utilities with nonnegative weights as

$$\rho(\mathscr{P}_{Sl2n}, \psi) = \sum_{i=1}^N \psi_i U_{S_i l2n}(P_{S_i l2n}, \mathscr{P}_{-S_i l2n}). \qquad (4.15)$$

The pseudo-gradient of $\rho(\mathscr{P}_{Sl2n}, \psi)$ is given as

$$\mathscr{H}(\mathscr{P}_{Sl2n}, \psi) = \left[\psi_1 \frac{U_{S_1 l2n}}{P_{S_1 l2n}}, \ldots, \psi_N \frac{U_{S_N l2n}}{P_{S_N l2n}}\right]^T. \qquad (4.16)$$

The Jacobian matrix, $\mathbf{H}(\mathscr{P}_{Sl2n}, \psi)$, of $\mathscr{H}(\mathscr{P}_{Sl2n}, \psi)$ w.r.t. $\mathscr{P}_{Sl2n}$ can be written as

$$H_{ij} = \begin{cases} \psi_i \dfrac{\partial^2 U_{S_i l 2n}}{\partial^2 P_{S_i l 2n}} < 0, & \text{for } i = j; \\[2mm] \psi_i \dfrac{\partial^2 U_{S_i l 2n}}{\partial P_{S_i l 2n} \partial P_{S_j l 2n}} = 0, & \text{for } i \neq j, \end{cases} \tag{4.17}$$

showing that $\mathbf{H}(\mathscr{P}_{Sl2n}, \psi)$ is a diagonal matrix with its all diagonal entries negative, thus $\mathbf{H}(\mathscr{P}_{Sl2n}, \psi)$ is negative definite. Therefore $[\mathbf{H}(\mathscr{P}_{Sl2n}, \psi) + \mathbf{H}^T(\mathscr{P}_{Sl2n}, \psi)]$ is negative definite as well. This indicates the weighted sum of utilities $\rho(\mathscr{P}_{Sl2n}, \psi)$ is diagonally strictly concave. According to Lemma 5, the NE is unique in the transmit power strategy $\mathscr{P}_{Sl2n}$ for the non-cooperative $AG = [\mathscr{S}, \mathscr{P}_S^{AG}, \mathscr{U}_S]$.

The NE in the sum-constrained joint power allocation game can be found through calculating the best response function for power allocation pair $\{P_{S_i l 2n}, P_{S_i P l 2n}\}$. Based on Karush-Kuhn-Tucker (KKT) conditions, the best response function is obtained by solving the following equation set

$$P_{S_i P l 2n} > 0, P_{S_i l 2n} > 0, \text{ and } \lambda_{ji} \geq 0, \text{ for } j = 1, 2, 3 \tag{4.18a}$$

$$\lambda_{1i}(P_{S_i l 2n} + P_{S_i P l 2n} - P_{S_i M}) = 0 \tag{4.18b}$$

(P4)
$$\lambda_{2i} P_{S_i P l 2n} = 0 \tag{4.18c}$$

$$\lambda_{3i} P_{S_i l 2n} = 0 \tag{4.18d}$$

$$\frac{\partial \mathscr{L}_{S_i l 2n}}{\partial P_{S_i P l 2n}} = 0 \text{ and } \frac{\partial \mathscr{L}_{S_i l 2n}}{\partial P_{S_i l 2n}} = 0 \tag{4.18e}$$

where the Lagrange function $\mathscr{L}_{S_i l 2n}$ is written as

$$\mathscr{L}_{S_i l 2n} = U_{S_i l 2n} + \lambda_{1i}(P_{S_i l 2n} + P_{S_i P l 2n} - P_{S_i M})$$
$$- \lambda_{2i} P_{S_i P l 2n} - \lambda_{3i} P_{S_i l 2n}$$

and $\lambda_{ji} \geq 0$ $(j = 1, 2, 3)$ are Lagrange multipliers. Due to space limitations and generalization of ϑ_i and ϵ_i, detailed calculation of KKT conditions is omitted in this chapter.

Finally, the unique NE $\{P^*_{S_i l 2n}, P^*_{S_i P l 2n}\}$ for the non-cooperative $AG = [\mathscr{S}, \mathscr{P}_S^{AG}, \mathscr{U}_S]$ is achieved by solving two equation sets, each consisting of N best response functions.

4.4.2 NE Analysis for CG

In this subsection, we analyze the NE for the non-cooperation three-phase $CG = [\mathscr{S}, \mathscr{P}_S^{CG}, \mathscr{U}_S]$.

Since SU_i transmits and relays in different time slots, $P_{S_i l3n}$ and $P_{S_i Pl3n}$ are independent, which implies they can be analyzed separately. Furthermore, $P_{S_i l3n}$ only affects SU_i's own utility. In this context, optimal solution to $P^*_{S_i l3n}$ is attained by SU_i's independent optimization. Therefore, only $P^*_{S_i l3n}$ needs to be addressed by game theoretic approach.

With the same analysis in AG, the throughput constraint can be removed by getting the cooperation region. Therefore, (P2) in three-phase relay can be equivalently rewritten as

$$\textbf{(P5)} \max_{P_{S_i l3n}, P_{S_i Pl3n}} U_{S_i l3n} = \frac{\epsilon_i P_{S_i Pl3n} C(P_{S_i l3n} \Gamma_{SS_i})}{3 \sum_{j=1}^{N} P_{S_j Pl3n}} \tag{4.19}$$

$$- \frac{\vartheta_i P_{S_i Pl3n}}{3}$$

$$\text{Subject to}: \ 0 < P_{S_i l3n}, P_{S_i Pl3n} \le P_{S_i}.$$

Theorem 4. *A NE in $\mathscr{P}_{S_i Pl3n}$ for the non-cooperative three-phase CG $=$ $[\mathscr{S}, \mathscr{P}_S^{CG}, \mathscr{U}_S]$ exists.*

Proof 8. According to the physical interpretation of $P_{S_i Pl3n} \in (0, P_{S_i M}]$, it is evident that $\mathscr{P}_{SPl3n}$ is a nonempty, convex and compact subset of an Euclidean space $\mathfrak{R}_+^N$. The second-order derivative w.r.t. $P_{S_i Pl3n}$ which is written as

$$\frac{\partial^2 U_{S_i l3n}}{\partial^2 P_{S_i Pl3n}} = - \frac{2\epsilon_i C(P_{S_i l3n} \Gamma_{SS_i}) \sum_{j \ne i} P_{S_j Pl3n}}{3 (\sum_{j=1}^{N} P_{S_j Pl3n})^3}$$

$$- \frac{\vartheta_i'' P_{S_i Pl3n} + 2\vartheta_i'}{3} \tag{4.20}$$

is always less than 0; thus $U_{S_i l3n}$ is concave in $\mathscr{P}_{S_i Pl3n}$.

Theorem 5. *A unique NE in $\mathscr{P}_{S_i Pl3n}$ exists for the non-cooperative CG $=$ $[\mathscr{S}, \mathscr{P}_S^{CG}, \mathscr{U}_S]$.*

Proof 9. To prove and analyze the uniqueness of the NE w.r.t. $P_{S_i Pl3n}$ for the non-cooperative $CG = [\mathscr{S}, \mathscr{P}_S^{CG}, \mathscr{U}_S]$, we calculate the best response function of SU_i by solving the equation when the first derivative of $U_{S_i l3n}$ w.r.t. $P_{S_i Pl3n}$ equating the result to 0, i.e.,

$$\frac{\partial U_{S_i l3n}}{\partial P_{S_i Pl3n}} = \underbrace{\frac{\epsilon_i C(P_{S_i l3n} \Gamma_{SS_i}) \sum_{j \ne i} P_{S_j Pl3n}}{(P_{S_i Pl3n} + \sum_{j \ne i} P_{S_j Pl3n})^2}}_{f_1(P_{S_i Pl3n})}$$

$$\underbrace{- (\vartheta_i' P_{S_i Pl3n} + \vartheta_i)}_{f_2(P_{S_i Pl3n})} = 0 \tag{4.21}$$

and we examine whether the best response function is a standard function which has the following three properties: positivity, monotonicity, and scalability [8].

It can be verified that $f_{1i}(P_{S_iPl3n})$ and the straight line $f_{2i}(P_{S_iPl3n})$ always has a unique intersection point for $P_{S_iPl3n} > 0$, when $f_{1i}(0) > f_{2i}(0)$ which yields $\epsilon_i C(P_{S_il3n}\Gamma_{SS_i}) > \vartheta_i \sum_{j\neq i} P_{S_jPl3n}$. This shows there is a unique solution to (4.21) for a feasible P_{S_iPl3n}. We denote the intersection point of $f_{1i}(P_{S_iPl3n})$ and the line $f_{2i}(P_{S_iPl3n})$ as $(\tilde{P}_{S_iPl3n}, A_i)$ and $A_i > \vartheta_i$, thus the best response function can be obtained by solving the following equation

$$\frac{\epsilon_i C(P_{S_il3n}\Gamma_{SS_i}) \sum_{j\neq i} P_{S_jPl3n}}{(\tilde{P}_{S_iPl3n} + \sum_{j\neq i} P_{S_jPl3n})^2} = A_i. \tag{4.22}$$

Eliminating the trivial cases in (4.22), i.e, the value which is less than 0 or exceeds the power limit P_{S_iM}, we can obtain SU_i's best response function $\Upsilon_i(\mathscr{P}_{SPl3n})$ as

$$\Upsilon_i = \sqrt{\frac{\epsilon_i C(P_{S_il3n}\Gamma_{SS_i}) \sum_{j\neq i} P_{S_jPl3n}}{A_i} - \sum_{j\neq i} P_{S_jPl3n}} \tag{4.23}$$

with the constraint

$$0 < \Upsilon_i(\mathscr{P}_{SPl3n}) \leq P_{S_iM} \tag{4.24}$$

when $\frac{\epsilon_i}{\vartheta_i} C(P_{S_il3n}\Gamma_{SS_i}) > \sum_{j\neq i} P_{S_jPl3n}, \forall S_i \in \mathscr{S}$.

Notice that $\Upsilon_i(\mathscr{P}_{SPl3n})$ is a quadratic function w.r.t. $\sqrt{\sum_{j\neq i} P_{S_jPl3n}}$, then $\Upsilon_i(\mathscr{P}_{SPl3n})$ is monotonically increasing w.r.t. $\sum_{j\neq i} P_{S_jPl3n}$ when $\sum_{j\neq i} P_{S_jPl3n} \leq \frac{\epsilon_i C(P_{S_il3n}\Gamma_{SS_i})}{4A_i}$. Since $\sum_{j\neq i} P_{S_jPl3n} \leq \frac{\epsilon_i C(P_{S_il3n}\Gamma_{SS_i})}{\vartheta_i}$, and $\frac{\epsilon_i C(P_{S_il3n}\Gamma_{SS_i})}{4A_i} < \frac{\epsilon_i C(P_{S_il3n}\Gamma_{SS_i})}{4\vartheta_i} < \frac{\epsilon_i C(P_{S_il3n}\Gamma_{SS_i})}{\vartheta_i}$, thus $\Upsilon_i(\mathscr{P}_{SPl3n})$ is always a monotonically increasing function w.r.t. $\sum_{j\neq i} P_{S_jPl3n}$.

Furthermore, for $\varphi > 1$, the following result holds

$$\varphi \Upsilon_i(\mathscr{P}_{SPl3n}) - \Upsilon_i(\varphi \mathscr{P}_{SPl3n}) =$$

$$= (\varphi - \sqrt{\varphi})\sqrt{\frac{\epsilon_i C(P_{S_il3n}\Gamma_{SS_i}) \sum_{j\neq i} P_{S_jPl3n}}{A_i}} > 0$$

Υ_i is scalable, because $\varphi \Upsilon_i(\mathscr{P}_{SPl3n}) > \Upsilon_i(\varphi \mathscr{P}_{SPl3n})$.

In conclusion, there is a unique NE exists in $\mathscr{P}_{S_iPl3n}$ for the non-cooperative three-phase $CG = [\mathscr{S}, \mathscr{P}_S^{CG}, \mathscr{U}_S]$.

By solving the equation set (4.23) consisting of N equations, the resulting relaying power for SU_i is

$$P^*_{S_iPl3n} = \frac{(N-1)(\sum_{j\in\mathscr{S}}\frac{C(P_{S_il3n}\Gamma_{SS_i})}{C(P_{S_jl3n}\Gamma_{SS_j})} - N + 1)\epsilon_i}{A_iC(P_{S_il3n}\Gamma_{SS_i})(\sum_{j\in\mathscr{S}}\frac{1}{C(P_{S_jl3n}\Gamma_{SS_j})})^2}. \tag{4.25}$$

Note that equation (4.25) is a generalized representation of equation (4.24) in [10] and equation (4.16) in [8], as can be found that the cost per unit transmission energy in [10] is a constant ω and the revenue per throughput is a constant 1 in [8].

To obtain the optimal P_{S_ilmn}, the subproblem can be solved by using convex optimization. Since $\epsilon'_i \leq 0$ and $\epsilon''_i \leq 0$, $\frac{\partial^2 U_{S_il3n}}{\partial^2 P_{S_il3n}} < 0$, i.e., U_{S_il3n} is concave w.r.t. P_{S_il3n}. For $\epsilon'_i < 0$, SU_i's optimal transmit power $P^*_{S_ilmn}$ is achieved by solving the following equation

$$\frac{\partial U_{S_il3n}}{\partial P_{S_il3n}} = \epsilon'_iC(P^*_{S_ilmn}\Gamma_{SS_i}) + \frac{\epsilon_i\Gamma_{SS_i}}{1 + P^*_{S_ilmn}} = 0 \tag{4.26}$$

and $P^*_{S_ilmn} = P_{S_iM}$ if $\epsilon'_i = 0$, since $\frac{\partial U_{S_il3n}}{\partial P_{S_il3n}} > 0$, U_{S_il3n} is increasing w.r.t P_{S_il3n}, the maximum U_{S_il3n} is achieved when $P^*_{S_il3n}$ is on the boundary.

It can be seen that $P^*_{S_iPl3n}$ is related to P_{S_il3n} while $P^*_{S_il3n}$ is independent of P_{S_iPl3n}; thus we calculate $P^*_{S_il3n}$ by using (4.26) and then calculate SU_i's achievable capacity and substitute it into (4.25). Meanwhile, $P^*_{S_iPl3n}$ is a function w.r.t $P^*_{S_jl3n}$ $(j \neq i, S_j \in \mathscr{S})$. To obtain $P^*_{S_iPl3n}$, each SU needs to send its achievable capacity under its optimal transmit power to the SAP. The SAP broadcasts $\sum_{j\in\mathscr{S}}\frac{1}{C(P^*_{S_jl3n}\Gamma_{SS_j})}$ to all SUs. Finally, SU_i's optimal transmit and relay powers can be attained in three-phase relay.

4.4.3 Maximizing PU's Utility

In this subsection, we maximize PU's utility in direct transmission and the utility obtained using cooperation based on the obtained NE in the non-cooperative game.

If no cooperation is triggered in CCRN, the PU transmits using direct transmission. In this scenario, the PU aims to optimize P_{PD} based on the following result.

Theorem 6. *The PU achieves the maximum U_{PD} when*

$$P^*_{PD} = \min\left\{\max(\frac{\Gamma_{PP} - J(E_{P0})}{J(E_{P0})\Gamma_{PP}}, \frac{e^{R_{PT}} - 1}{\Gamma_{PP}}), P_{PM}\right\} \tag{4.27}$$

for $J(E_{P0}) > 0$; and the PU achieves the maximum U^*_{PD} when $P^*_{PD} = P_{PM}$ for $J(E_{P0}) = 0$, respectively.

Proof 10. In case that $J(E_{P0}) > 0$, U_{PD}^* is achieved when the first derivative of U_{PD} w.r.t. P_{PD}, i.e., $\frac{\partial U_{PD}}{\partial P_{PD}} = \frac{\Gamma_{PP}}{1+\Gamma_{PP}P_{PD}} - J(E_{P0})$, is 0. Meanwhile, $R_{PD} \geq R_{PT}$ yields $P_{PD} \geq \frac{e^{R_{PT}}-1}{\Gamma_{PP}}$. Due to the power constraint of PU, we obtain (4.27). For $J(E_{P0}) = 0$, as U_{PD} is increasing w.r.t P_{PD}, U_{PD}^* is always achieved at the boundary, i.e., $P_{PD}^* = P_{PM}$.

In the scenario of PUs cooperating with SUs, the optimization problem in (4.5) belongs to a type of nonconvex and nonlinear programming. There are eight possible communication strategies in total, thus PU's maximum utility for each communication strategy and then the optimal solution to (4.5) can be obtained by choosing the maximum utility among these eight communication strategies.

To obtain PU's maximum utility, the optimization problem is addressed by discussing AF and DF separately as the throughput under AF and DF has different properties.

In the AF forwarding mode, denote PU's energy consumption as $E_{PlmA} = W_{lmA}P_{PlmA}$, where W_{lmA} is PU's power consumption factor determined by the transmission and relay strategies as shown in Fig. 4.3, e.g., $W_{M3A} = \frac{1}{3}$ and $W_{C2A} = 1$. Denote the achievable capacity of the PU as $C_{PlmA} = mR_{PlmA}$.

Theorem 7. *In the AF forwarding strategy, for $J(E_{P0}) > 0$, the PU can achieve the maximum utility if P_{PlmA}^* is the solution of the following equation*

$$\frac{\partial C_{PlmA}}{\partial P_{PlmA}} = mW_{lmA}J(E_{P0}); \tag{4.28}$$

and for $J(E_{P0}) = 0$, $P_{PlmA}^* = P_{PM}$.

Proof 11. The first derivative of U_{PlmA} w.r.t P_{PlmA} is

$$\frac{\partial U_{PlmA}}{\partial P_{PlmA}} = \frac{1}{m}\frac{\partial C_{PlmA}}{\partial P_{PlmA}} - J(E_{P0})W_{lmA}. \tag{4.29}$$

Since $\frac{\partial C_{PlmA}}{\partial P_{PlmA}} > 0$ for all $l \in \{M, C\}$ and $m \in \{2, 3\}$, the maximum U_{PlmA} is achieved when its first derivative w.r.t. P_{PlmA} equals to 0, which yields (4.28). In particular, if $J(E_{P0}) = 0$, $\frac{\partial U_{PlmA}}{\partial P_{PlmA}} > 0$ showing U_{PlmA} is increasing w.r.t. P_{PlmA}, thus the maximum U_{PlmA} is achieved when P_{PlmA}^* is at the boundary, i.e., PU's power limit P_{PM}.

In the DF forwarding mode, denote PU's energy consumption as $E_{PlmD} = W_{lmD}P_{PlmD}$, where W_{lmD} is PU's power consumption factor determined by the transmission and relay strategies. We represent Γ_{PS} as the minimum channel gain among all the PU-SU links, i.e., $\Gamma_{PS} = \min_{S_i \in \mathscr{S}} \Gamma_{PS_i}$.

Theorem 8. *In cooperative DF, if $\Gamma_{PS} > \Gamma_{PP}$, the PU can achieve the maximum utility when*

$$P_{PCmD}^* = \frac{\sum_{i=1}^{N} P_{S_iPCmD}\Gamma_{S_iP}}{\Gamma_{PS} - \Gamma_{PP}}; \tag{4.30}$$

if $\Gamma_{PS} \leq \Gamma_{PP}$, the maximum utility is achieved when

$$P^*_{PCmD} = \left[\frac{1}{mW_{CmD}J(E_{P0})} - \frac{1}{\Gamma_{PS}} \right]^+ \tag{4.31}$$

for $J(E_{P0}) > 0$; and $P^*_{PCmD} = P_{PM}$ for $J(E_{P0}) = 0$.
   In multihop DF, the PU can achieve the maximum utility if

$$P^*_{PMmD} = \frac{\sum_{i=1}^{N} P_{S_iPCmD}\Gamma_{S_iP}}{\Gamma_{PS}}. \tag{4.32}$$

Proof 12. In the scenario of cooperative DF with $\Gamma_{PS} > \Gamma_{PP}$, since U_{PCmD} is increasing w.r.t. R_{PCmD}, and the maximum R_{PCmD} is achieved when the throughput of PU-SU link equals to that of {PU, SU}-PBS link, thus (4.30) is obtained. In case that $\Gamma_{PS} \leq \Gamma_{PP}$, the throughput of PU-SU link is always smaller than that of {PU, SU}-PBS link, hence $mR_{PCmD} = C(P_{PCmD}\Gamma_{PS})$. For $J(E_{P0}) > 0$, P^*_{PCmD} is obtained by solving $\frac{\partial R_{PCmD}}{\partial P_{PCmD}} = mJ(E_{P0})W_{CmD}$ which yields (4.31). For $J(E_{P0}) = 0$, P^*_{PCmD} is achieved at the boundary due to monotonicity of U_{PCmD} w.r.t. P_{PCmD}.

   In multihop DF, the maximum U_{PMmD} is achieved when R_{PMmD} is maximal, i.e., the throughput of PU-SU link equals to that of {PU, SU}-PBS link which yields (4.32).

4.4.4   Implementation Protocol

We consider that the SAP is fully trusted by the primary network, and it is reliable to the PU and fair to all SUs. The SAP measures Γ_{SS_k}. In case that Γ_{SS_k} is below a certain threshold, SU_k is impossible to achieve a satisfactory throughput. Thus the SAP selects those SUs with fading coefficients no less than predefined thresholds as potential candidate SUs. In addition, the SAP periodically collects Γ_{PP} and Γ_{S_kP} from the PBS, and Γ_{PS_k} is collected from SU_k. According to the above illustration, the SAP enumerates all possible $\mathscr{S}$ for each transmission-relay-forwarding strategy, which satisfy the corresponding cooperative set selection criteria, e.g., the constraint in (4.24) for three-phase relay. Based on each possible set $\mathscr{S}$, the SAP helps the PU calculate PU's optimal transmit power to obtain PU maximum utility. From all possible sets, the one that maximizes PU's utility function is selected to be the optimal relay set for that transmission-relay-forwarding strategy, and optimal PU power is set to be the corresponding parameter under the optimal cooperating set, e.g., according to (4.32) in multihop DF. After obtaining the optimal parameter for that transmission-relay-forwarding strategy, the SAP informs the selected SUs about P^*_{Plmn}. Then each SU calculates optimal powers for relaying and transmission individually, e.g., according to the KKT conditions in (4.18) for two-phase relay and (4.25) for three-phase relay. Since the optimal power calculation depends on $\sum_{j \in \mathscr{S}} \frac{1}{C(P_{S_{jl3n}})}$, the SAP also piggybacks $\sum_{j \in \mathscr{S}} \frac{1}{C(P_{S_{jl3n}})}$ to each SU.

As there are eight different transmission-relay-forwarding combining strategies, the SAP selects PU's highest maximum utility amongst the maximum utility of each communication strategy. The SAP sends the optimal decision on l^*, m^*, n^*, and P^*_{Plmn} to the PU and the SUs in the corresponding set $\mathscr{S}_{l^*m^*n^*}$. Finally, upon the optimal cooperative strategy from the SAP, the PU compares $U_{l^*m^*n^*}$ with U^*_{PD}, and informs the secondary network its final decision. In Sect. 4.5, two numerical results are given to unveil the impact of PU's energy

4.5 Performance Evaluation

In this section, two numerical results are given to evaluate the impact of PU's energy level and number of SUs on the optimal communication strategy selection. Rayleigh block fading is considered as the channel model, i.e., channel gains are invariant within a cooperation duration, and only the small-scale fading component is taken into account for simplicity. In our simulation, $J(E_{P0}) = \frac{\exp(50-E_{P0})}{10}$, $\vartheta = 10$, and $\epsilon = 5$ are used, and we set $P_{SM} = 1$ and $P_{PM} = 2$.

Figure 4.4a shows the impact of E_{P0} on PU's optimal communication strategy, where there are three relaying SUs. With the increase of E_{P0}, PU's utilities in logarithmic form increase for all communication strategies, as $J(E_{P0})$ decreases w.r.t E_{P0}. It is shown that with the same channel fading, PU's optimal strategy is related with E_{P0}. For example, PU's optimal strategy is achieved when $l^* = M$, $m^* = 2$, and $n^* = D$ if $E_{P0} = 52$, while $l^* = C$, $m^* = 3$, and $n^* = D$ if $E_{P0} = 56$. On the one hand, with a high energy level, PU's cost in energy consumption of direct transmission decreases. On the other hand, the PU can occupy the whole time duration for its own transmission in direct transmission, i.e., there is no pre-log factor in PU's throughput. Therefore, with the increase of E_{P0}, it can be seen that the difference between utilities of direct transmission and utilities in cooperation decreases.

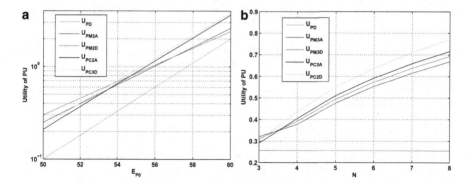

Fig. 4.4 PU utilities versus E_{P0} and N. (**a**) PU's utility of different energy level E_{P0}. (**b**) PU's utility of different number of SUs

Figure 4.4b shows the value of PU's utilities of different strategies under various numbers of relaying SUs, and PU's energy level is $E_{P0} = 56$. When N increases from 3 to 8, PU's utilities in all strategies increase due to the improved throughput performance. It is evident that the U_{PD} is a constant and is independent of the value of N.

4.6   Conclusion

This chapter presents a new framework to address the optimal communication strategy selection and the associated parameter optimization for maximizing PUs' utilities in CCRN. By taking the impacts of energy levels and power consumptions on communication strategies into consideration, the problem is formulated as two types of Stackelberg games, i.e., a sum-constrained joint power allocation game for two-phase cooperation, and a power control game for three-phase cooperation, respectively. The unique NE for each game is proved, and an implementation protocol is proposed to address partner selection and parameter optimization based on the obtained NE. Simulation results show that our scheme to achieve spectrum-energy efficient CCRN is viable and promising.

References

1. A. Goldsmith, S. A. Jafar, I. Maric, and S. Srinivasa, "Breaking spectrum gridlock with cognitive radios: An information theoretic perspective," *Proc. IEEE*, vol. 97, no. 5, pp. 894–914, May 2009.
2. Y.–C. Liang, K.–C. Cheng, G. Y. Li, and P. Mähönen, "Cognitive radio networking and communications: An overview," *IEEE Trans. Veh. Techno.*, vol. 60, no. 7, pp. 3386–3407, Sept. 2011.
3. Y. Liu, L. X. Cai, and X. Shen, "Spectrum-aware opportunistic routing in multi-hop cognitive radio networks," *IEEE J. Sel. Areas Commun.*, to appear.
4. A. Alshamrani, X. Shen, and L. Xie, "QoS provisioning for heterogeneous services in cooperative cognitive radio networks," *IEEE J. Sel. Areas Commun.* vol. 29, no. 4, pp. 819–830, Apr. 2011.
5. H. T. Cheng, W. Zhuang, "Simple channel sensing order in cognitive radio networks," *IEEE J. Sel. Areas Commun.*, no. 29, vol. 4, pp. 676–688, Apr. 2011.
6. B. Cao, Q. Zhang, J. W. Mark, L. X. Cai, and H. V. Poor, "Toward efficient radio spectrum utilization: User cooperation in cognitive radio networking," *IEEE Network*, vol. 26, no. 4, pp. 46–52, Jul. 2012.
7. O. Simeone, I. Stanojev, S. Savazzi, Y. Bar-Ness, U. Spagnolini, and R. Pickholtz, "Spectrum leasing to cooperating secondary ad hoc networks," *IEEE J. Sel. Areas Commun.*, vol. 26, no. 1, pp. 203–213, Jan. 2008.
8. J. Zhang and Q. Zhang, "Stackelberg game for utility-based cooperative cognitive radio networks," in *Proc. MobiHoc*, pp. 23–32, New Orleans, LA., May 2009.
9. W. Su, J. D. Matyjas, and S. Batalama, "Active cooperation between primary users and cognitive radio users in cognitive ad-hoc networks," *IEEE Trans. Signal Proc.*, vol. 60, no. 4, pp. 1796–1805, Apr. 2012.

10. S. Hua, H. Liu, M. Wu, and S. Panwar, "Exploiting MIMO antennas in cooperative cognitive radio networks," in *Proc. INFOCOM*, pp. 2714–2722, Shanghai, China, Apr. 2011.
11. B. Cao, L. X. Cai, H. Liang, J. W. Mark, Q. Zhang, H. V. Poor, and W. Zhuang, "Cooperative cognitive radio networking using quadrature signaling," in *Proc. INFOCOM*, pp. 3096–3100, Orlando, FL., Mar. 2012.
12. H. Xu and B. Li, "Efficient resource allocation with flexible channel cooperation in OFDMA cognitive radio networks," in *Proc. INFOCOM*, pp. 1–9, San Diego, CA., Mar. 2010.
13. X. Hao, M. H. Cheung, V. W. Wong, and V. C. Leung, "A Stackelberg game for cooperative transmission and random access in cognitive radio networks," in *IEEE PIMRC*, pp. 411–416, Toronto, Canada, Sept. 2011.
14. B. Cao, J. W. Mark, and Q. Zhang, "A polarization enabled cooperation framework for cognitive radio networking," in *Proc. GLOBECOM*, Anaheim, CA., Dec. 2012.
15. Y. Su and M. van der Schaar, "Additively Coupled Sum Constrained Games" *Proc. GameNets* Shanghai, China, Apr. 2011.
16. J. Rosen, "Existence and uniqueness of equilibrium points for concave N-person games," *Econometrica*, vol. 33, no. 3, pp. 520–534, 1965.
17. J. Goodman, "A note on existence and uniqueness of equilibrium points for concave N-person games," *Econometrica*, vol. 48, no. 1, pp. 251, 1980.

Chapter 5
Conclusions and Closing Remarks

With the rapid progress in communication technologies and the explosive proliferation of wireless applications, the demand for wireless broadband services continues to explode. On the one hand, as a precious natural resource, the amount of most easily explorable radio spectrum for wireless communications is extremely limited. On the other hand, the significant spectrum underutilization resulting from current fixed spectrum allocation polices has even exacerbated the situation of spectrum scarcity. In order to address the spectrum underutilization arising from traditional static spectrum allocation polices, advanced modifications and improvements in spectrum access & management mechanisms are promising to enhance spectrum utilization, among them dynamic spectrum access (DSA) is notable. In the classical DSA framework, unlicensed spectrum users, also known as secondary users (SUs), are enabled to access to the licensed spectrum for wireless service transmissions, as long as their communications cause no (or acceptable) harm to licensed spectrum users, also known as primary users (PUs). Meanwhile, prorogation mediums of wireless communications are time varying and become more and more complicated, such as severe multipath fading in dense buildings and fast moving terminals, leading to considerable low spectrum efficiency. Due to spectrum shortage, it is now difficult to improve or even guarantee a satisfied quality of service (QoS) by leveraging large bandwidth. As a promising paradigm to improve communication performance of wireless systems and networks, user cooperation frameworks and their related techniques can be directly applied for existing technologies and mechanisms to effectively enhance spectrum efficiency. Therefore, in user cooperation enabled DSA systems, SUs are allowed to access to licensed spectrum via actively helping with PUs' transmissions. In such a way, spectrum utilization is improved through DSA, and spectrum efficiency is enhanced by user cooperation.

By leveraging the advantages of DSA in spectrum utilization and the advantages of user cooperation in spectrum efficiency, this monograph has aimed to investigate the system model and its critical techniques to engineering the user cooperation enabled DSA systems. In such a system, DSA allows SUs to improve spectrum utilization via accessing to licensed spectrum, and user cooperation techniques enable PUs to improve communication performance via enhancing spectrum efficiencies, i.e., creating a win-win situation. To achieve this goal, the feasibility of introducing cooperation between PUs and SUs into DSA system is first studied. Technique characteristics and performance metrics of user cooperation enabled DSA for classical application scenario are discussed. Based on state-of-the-art signal processing, transceivers design, user cooperation, and cooperation scenarios, two-phase cooperative relay & transmission models and frame structures are conducted. In addition, associated signal processing techniques and algorithms are proposed to mitigate the co-channel interference resulting from concurrent relay & transmission. To further investigate the performance of the proposed models, by taking limited and constrained resources at networks and terminals into consideration, performance metrics in terms of objective and utility functions and their corresponding solutions and algorithms are formulated and given to perform the optimal resource management and allocation. Based on the aforementioned contents, classical transmission-forward-relay strategies are summarised to study the optimal communication strategy problem in user cooperation enabled DSA. The main aims of this monograph can be concluded as follows.

1. System modeling and analysis of user cooperation enabled DSA. Based on the observation of PUs' requirement in communication performance and SUs' requirement in spectrum access, the rationality of combining these two requirements by leveraging the advantages of DSA in spectrum utilization and the advantages of user cooperation in spectrum efficiency is discussed. Moreover, the feasibility of user cooperation techniques and policy issues is analyzed. A cooperative relay & transmission model and secondary spectrum leasing model and their critical issues are proposed to enable the applications of user cooperation based DSA in different scenarios. Focusing on the cooperative relay & transmission model, the throughput and energy consumption performance of a three-phase cooperative relay & transmission scheme and a two-phase cooperative relay & transmission scheme with co-channel interference are studied and compared. Some common-used service scenarios and mathematical models are summarized.

2. System modeling and analysis of two-phase cooperative relay & transmission models. In order to improve the cooperation efficiency, two-phase cooperative relay & transmission models, their frame structures and user selection protocols are proposed. To mitigate the co-channel interference arising from concurrent relay & transmission in single antenna systems, a quadrature modulation enabled two-phase cooperation framework is studied. In addition, the associated transceiver design and signaling plan are presented to investigate the cooperation mechanisms. To mitigate the co-channel interference in multi-antenna systems, an orthogonally dual-polarized antenna (ODPA) based two-phase cooperation framework is proposed. Based on polarization signal processing and oblique

projection theory, oblique projection polarization filtering, polarization alignment, and polarization zero-forcing are performed to unveil the cooperation mechanisms.

3. Resource management and optimization in two-phase cooperative relay & transmission models. By bearing the capability of network and user, communication efficiency and reliability in mind, the optimal resource allocation mechanisms and strategies are analyzed. Specifically, in quadrature modulation based models with different relaying modes, convex programming is used to conduct the maximization of weighted sum throughput of one PU and one SU, and the optimal powers and time fractions are obtained in closed-form. Stackelberg game theory is utilized to address the optimal power allocations in the cooperation between one PU and multiple SUs. In the ODPA based models, geometric programming and Signomial programming are used to tackle the sum throughput maximization of one PU and one SU; Markov decision process, dynamic programming, heuristic online algorithm, and approximation algorithm are used to conduct the analysis on the optimal power polices between one PU and two SUs. By approximating the original nonconvex programming to a convex one, convex programming is utilized to devise a centralized and a distributed algorithms for the sum throughput maximization of one PU with multiple SUs.

4. The optimal communication strategy design and analysis on cooperative relay & transmission between PUs and SUs. The issues of whether to cooperate and how to cooperate are discussed. In the non-cooperative scenario, the optimal resource management and allocation for PUs are studied. By respectively combining three-phase and two-phase cooperative models with different forwarding and relaying strategies, nine common-used communication models and their communication efficiency are analysed. To qualitatively reflect the cooperation requirements of PUs and SUs, spectrum- and energy-efficiency based utility functions are formulated. According to different features in three-phase models and two-phase models, non-cooperative game theory is used to study the optimal power strategies of PUs and SUs. In specific, a non-cooperative power allocation game is established for the three-phase model, and a non-cooperation power control game is used for the two-phase model. The existence and uniqueness of Nash equilibria (NE) for both games are proved, and NE points are respectively provided in analytical format. Given the energy, power, and throughput requirements, this result can provide the optimal transmission-forward-relay strategies to achieve the maximum utilities for both PUs and SUs.

Printed in the United States
By Bookmasters